ENERGY-EFFICIENT ELECTRIC MOTORS

ELECTRICAL AND COMPUTER ENGINEERING
A Series of Reference Books and Textbooks

FOUNDING EDITOR

Marlin O. Thurston
Department of Electrical Engineering
The Ohio State University
Columbus, Ohio

ENERGY-EFFICIENT ELECTRIC MOTORS

Third Edition, Revised and Expanded

ALI EMADI
Illinois Institute of Technology
Chicago, Illinois

MARCEL DEKKER, INC.　　　　　　　　NEW YORK

Previous edition titled *Energy-Efficient Electric Motors: Selection and Application, Second Edition*, John C. Andreas, Marcel Dekker, 1992.

Library of Congress Cataloging-in-Publication Data
A catalog record for this book is available from the Library of Congress.

ISBN: 0-8247-5735-1

This book is printed on acid-free paper.

Headquarters
Marcel Dekker, 270 Madison Avenue, New York, NY 10016, U.S.A.
tel: 212-696-9000; fax: 212-685-4540

Distribution and Customer Service
Marcel Dekker, Cimarron Road, Monticello, New York, NY 12701, U.S.A.
tel: 800-228-1160; fax: 845-796-1772

World Wide Web
http://www.dekker.com

The publisher offers discounts on this book when ordered in bulk quantities. For more information, write to Special Sales/Professional Marketing at the headquarters address above.

Current printing (last digit):

10   9   8   7   6   5   4   3   2   1

PRINTED IN THE UNITED STATES OF AMERICA

To John C. Andreas

.

Preface

The main purpose of this new edition continues to be to provide guidelines for selecting and utilizing electric motors on the basis of energy efficiency and life-cycle cost. In previous editions of this book, particular emphasis was given to three-phase and single-phase induction motors in the 1–200 hp range since this was the range offering maximum opportunities for energy savings. However, since the second edition, there has been a growing demand in the direction of solid-state intensive electric motor drives as adjustable or variable speed drives. New electric motors such as brushless DC and switched reluctance have also been mass-produced and made commercially available. The impetus toward this expansion of power electronics has been provided by recent advancements in the areas of solid-state switching devices, control electronics, and advanced microcontrollers, microprocessors, and digital signal processors (DSP). These advancements facilitate high-tech applications and enable the introduction of power electronic converters with highest performance, maximum efficiency, and minimum volume and weight. In fact, electric motors with advanced power electronic drivers have real and significant potential for improving not only efficiency and life-cycle cost, but also reliability, performance, and safety.

In this edition, Chapters 1, 2, 4, 5, and 7 from the previous editions have been updated, rearranged, and revised. These chapters present energy-efficient single-phase and three-phase induction motors comprehensively. Chapters 3, 6, 8, 9, and 10 are new. Chapter 3 presents the fundamentals of power electronics applicable to electric motor drives. Adjustable speed drives and their applications are explained in Chapter 6. Advanced permanent magnet (PM) and brushless DC (BLDC) motor drives as well as switched reluctance motor (SRM) drives are presented in Chapters 8 and 9, respectively. Finally, utility interface issues including power factor correction (PFC) and active filters (AF) are discussed in Chapter 10.

I would like to acknowledge gratefully the contributions of many graduate students at Illinois Institute of Technology in different sections/chapters of this book. They are Mr. Brian Kaczor contributing in Chapter 3, Mr. Timothy R. Cooke, Mr. Anthony Villagomez, and Mr. Semih Aslan contributing in Chapter 6, Mr. Manas C. Phadke and Mr. Aly A. Aboul-Naga contributing in Chapter 8, Mr. Himanshu Ray, Ms. Alpa Bhesania, Mr. Madan M. Jalla, Mr. Sheldon S. Williamson, Mr. Piyush C. Desai, and Mr. Ranjit Jayabalan contributing in Chapter 9, and Mr. Ritesh Oza and Mr. Abdolhosein Nasiri contributing in Chapter 10.

I would also like to acknowledge the efforts and assistance of the staff of Marcel Dekker, Inc.

Ali Emadi

Contents

1

Induction Motor Characteristics

1.1 THREE-PHASE INDUCTION MOTORS

In the integral horsepower sizes, i.e., above 1 hp, three-phase induction motors of various types drive more industrial equipment than any other means. The most common three-phase (polyphase) induction motors fall within the following major types:

NEMA (*National Electrical Manufacturers Association*) design B: Normal torques, normal slip, normal locked amperes

NEMA design A: High torques, low slip, high locked amperes

NEMA design C: High torques, normal slip, normal locked amperes

NEMA design D: High locked-rotor torque, high slip

Wound-rotor: Characteristics depend on external resistance

Multispeed: Characteristics depend on design—variable
torque, constant torque, constant horsepower

There are many specially designed electric motors with unique
characteristics to meet specific needs. However, the majority of
needs can be met with the preceding motors.

1.1.1   NEMA Design B Motors

The NEMA design B motor is the basic integral horsepower motor.
It is a three-phase motor designed with normal torque and normal
starting current and generally has a slip at the rated load of less than
4%. Thus, the motor speed in revolutions per minute is 96% or more
of the synchronous speed for the motor. For example, a four-pole
motor operating on a 60-Hz line frequency has a synchronous speed
of 1800 rpm or a full-load speed of

$$1800 - (1800 \times \text{slip}) = 1800 - (1800 \times 0.04)$$

$$= 1800 - 72$$

$$= 1728 \text{ rpm}$$

or

$$1800 \times 0.96 = 1728 \text{ rpm}$$

In general, most three-phase motors in the 1- to 200-hp range have a
slip at the rated load of approximately 3% or, in the case of four-
pole motors, a full-load speed of 1745 rpm. Figure 1.1 shows the
typical construction for a totally enclosed, fan-cooled NEMA
design B motor with a die-cast aluminum single-cage rotor.

Figure 1.2 shows the typical speed-torque curve for the NEMA
design B motor. This type of motor has moderate starting torque, a
pull-up torque exceeding the full-load torque, and a breakdown
torque (or maximum torque) several times the full-load torque.
Thus, it can provide starting and smooth acceleration for most loads
and, in addition, can sustain temporary peak loads without stalling.
The NEMA performance standards for design B motors are shown
in Tables 1.1–1.3.

FIGURE 1.1 NEMA design B totally enclosed, fan-cooled polyphase induction motor. (Courtesy Magnetek, St. Louis, MO.)

In the past, there were no established standards for efficiency or power factor for NEMA design B induction motors. However, NEMA had established standards for testing and labeling induction motors. Recently, NEMA has established efficiency standards for energy-efficient polyphase induction motors. These standards are discussed in detail in Chapter 2.

1.1.2 NEMA Design A Motors

The NEMA design A motor is a polyphase, squirrel-cage induction motor designed with torques and locked-rotor current that exceed the corresponding values for NEMA design B motors. The criterion for classification as a design A motor is that the value of the locked-rotor current be in excess of the value for NEMA design B motors. The NEMA design A motor is usually applied to special applications that cannot be served by NEMA design B motors, and most often these applications require motors with higher than normal breakdown torques to meet the requirements of high transient or short-duration loads. The NEMA design A motor is also applied to loads requiring extremely low slip, on the order of 1% or less.

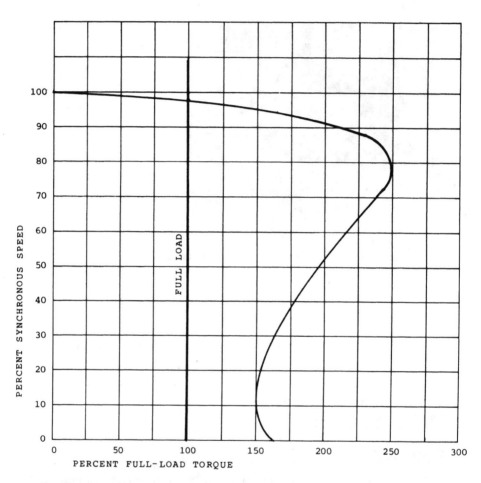

FIGURE 1.2   NEMA design B motor speed-torque curve.

1.1.3 NEMA Design C Motors

The NEMA design C motors is a squirrel-cage induction motor that develops high locked-rotor torques for hard-to-start applications. Figure 1.3 shows the construction of a drip-proof NEMA design C motor with a double-cage, die-cast aluminum rotor. Figure 1.4 shows the typical speed torque curve for the NEMA design C motor. These motors have a slip at the rated load of less than 5%.

TABLE 1.1 Locked-Rotor Torque of NEMA Design A and B Motors[a,b]

hp	Synchronous speed, 60 Hz			
	3600 rpm	1800 rpm	1200 rpm	900 rpm
1	—	275	170	135
1.5	175	250	165	130
2	170	235	160	130
3	160	215	155	130
5	150	185	150	130
7.5	140	175	150	125
10	135	165	150	120
15	130	160	140	125
20	130	150	135	125
25	130	150	135	125
30	130	150	135	125
40	125	140	135	125
50	120	140	135	125
60	120	140	135	125
75	105	140	135	125
100	105	125	125	125
125	100	110	125	120
150	100	110	120	120
200	100	100	120	120
250	70	80	100	100

[a] Single-speed, polyphase, squirrel-cage, medium-horsepower motors with continuous ratings (percent of full-load torque).
[b] For other speeds and ratings, see NEMA Standard MG1-12.38.1.
Source: Reprinted by permission from NEMA Standards Publication No. MG1-1987 Motor and Generators, copyright 1987 by the National Electrical Manufacturers Association.

The NEMA performance standards for NEMA design C motors are shown in Tables 1.3–1.5.

1.1.4 NEMA Design D Motors

The NEMA design D motor combines high locked-rotor torque with high full-load slip. Two standard designs are generally offered,

TABLE 1.2 Breakdown Torque of NEMA Design A and B Motors[a,b]

	Synchronous speed, 60 Hz			
hp	3600 rpm	1800 rpm	1200 rpm	900 rpm
1	—	300	265	215
1.5	250	280	250	210
2	240	270	240	210
3	230	250	230	205
5	215	225	215	205
7.5	200	215	205	200
10	200	200	200	200
15	200	200	200	200
20	200	200	200	200
25	200	200	200	200
30	200	200	200	200
40	200	200	200	200
50	200	200	200	200
60	200	200	200	200
75	200	200	200	200
100	200	200	200	200
125	200	200	200	200
150	200	200	200	200
200	200	200	200	200
250	175	175	175	175

[a] Single-speed, polyphase, squirrel-cage, medium-horsepower motors with continuous ratings (percent of full-load torque).
[b] For other speeds and ratings, see NEMA Standard MG1-12.39.1.
Source: Reprinted by permission from NEMA Standards Publication No. MG1-1987 Motors and Generators, copyright 1987 by the National Electrical Manufacturers Association.

one with full-load slip of 5–8% and the other with full-load slip of 8–13%. The locked-rotor torque for both types is generally 275–300% of full-load torque; however, for special applications, the locked-rotor torque can be higher. Figure 1.5 shows the typical speed-torque curves for NEMA design D motors. These motors are recommended for cyclical loads such as those found in punch

TABLE 1.3 Locked-Rotor Current of NEMA Design B, C, and D Motors[a,b,c]

hp	Locked-rotor current A	NEMA design letter	Code letter
1	30	B, D	N
1.5	40	B, D	M
2	50	B, D	L
3	64	B, C, D	K
5	92	B, C, D	J
7.5	127	B, C, D	H
10	162	B, C, D	H
15	232	B, C, D	G
20	290	B, C, D	G
25	365	B, C, D	G
30	435	B, C, D	G
40	580	B, C, D	G
50	725	B, C, D	G
60	870	B, C, D	G
75	1085	B, C, D	G
100	1450	B, C, D	G
125	1815	B, C, D	G
150	2170	B, C, D	G
200	2900	B, C	G
250	3650	B	G

[a] Three-phase, 60-Hz, medium-horsepower, squirrel-cage induction motors rated at 230 V.

[b] For other horsepower ratings, see NEMA Standard MG1-12.35.

[c] The locked-rotor current for motors designed for voltages other than 230 V shall be inversely proportional to the voltage.

Source: Reprinted by permission from NEMA Standards Publication No. MG1-1987, Motors and Generators, copyright 1987 by the National Electrical Manufacturers Association.

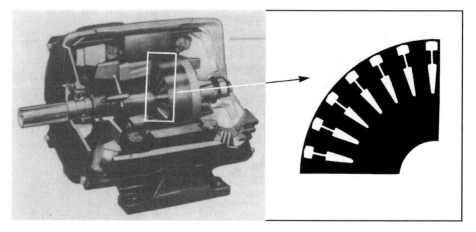

FIGURE 1.3 NEMA design C drip-proof polyphase induction motor. (Courtesy Magnetek, St. Louis, MO.)

presses, which have stored energy systems in the form of flywheels to average the motor load and are excellent for loads of short duration with frequent starts and stops. The proper application of this type of motor requires detailed information about the system inertia, duty cycle, and operating load as well as the motor characteristics. With this information, the motors are selected and applied on the basis of their thermal capacity.

1.1.5 Wound-Rotor Induction Motors

The wound-rotor induction motor is an induction motor in which the secondary (or rotating) winding is an insulated polyphase winding similar to the stator winding. The rotor winding generally terminates at collector rings on the rotor, and stationary brushes are in contact with each collector ring to provide access to the rotor circuit. A number of systems are available to control the secondary resistance of the motor and hence the motor's characterstics. The use and application of wound-rotor induction motors have been limited mostly to hoist and crane applications and special speed-

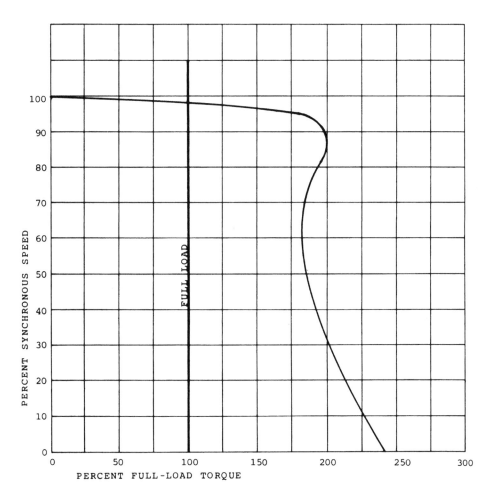

FIGURE 1.4 NEMA design C motor speed-torque curve.

control applications. Typical wound-rotor motor speed-torque curves for various values of resistance inserted in the rotor circuit are shown in Fig. 1.6. As the value of resistance is increased, the characteristic of the speed-torque curve progresses from curve 1 with no external resistance to curve 4 with high external resistance. With appropriate control equipment, the characteristics of the

TABLE 1.4 Locked-Rotor Torque of NEMA Design C Motors[a]

	Synchronous speed, 60 Hz		
hp	1800 rpm	1200 rpm	900 rpm
3	—	250	225
5	250	250	225
7.5	250	225	200
10	250	225	200
15	225	200	200
20–200 inclusive	200	200	200

[a] Single-speed, polyphase, squirrel-cage, medium-horsepower motors with continuous ratings (percent of full-load torque), MG1-12.38.2.
Source: Reprinted by permission from NEMA Standards Publication No. MG1-1987, Motors and Generators, copyright 1987 by the National Electrical Manufacturers Association.

TABLE 1.5 Breakdown Torque of NEMA Design C Motors[a]

	Synchronous speed, 60 Hz		
hp	1800 rpm	1200 rpm	900 rpm
3	—	225	200
5	200	200	200
7.5–200 inclusive	190	190	190

[a] Single-speed, polyphase, squirrel-cage, medium-horsepower motors with continuous ratings (percent of full-load torque), MG1-12.39.2.
Source: Reprinted by permission from NEMA Standards Publication No. MG1-1987, Motors and Generators, copyright 1987 by the National Electrical Manufacturers Association.

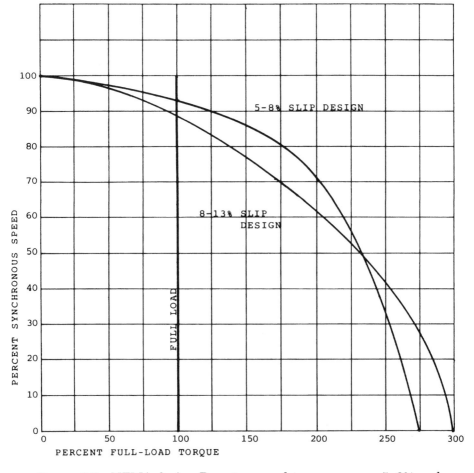

FIGURE 1.5 NEMA design D motor speed-torque curves: 5–8% and 8–13% slip.

motor can be changed by changing this value of external rotor resistance. Solid-state inverter systems have been developed that, when connected in the rotor circuit instead of resistors, return the slip loss of the motor to the power line. This system substantially improves the efficiency of the wound-rotor motor used in variable-speed applications.

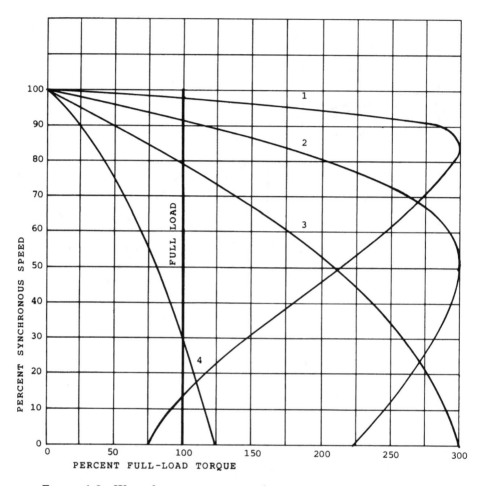

FIGURE 1.6 Wound-rotor motor speed-torque curves: 1, rotor short-circuited; 2–4, increasing values of external resistance.

1.1.6 Multispeed Motors

Motors that operate at more than one speed, with characteristics similar to those of the NEMA-type single-speed motors, are also available. The multispeed induction motors usually have one or two primary windings. In one-winding motors, the ratio of the two speeds must be 2 to 1; for example, possible speed combinations are

3600/1800, 1800/900, and 1200/600 rpm. In two-winding motors, the ratio of the speeds can be any combination within certain design limits, depending on the number of winding slots in the stator. The most popular combinations are 1800/1200, 1800/900, and 1800/600 rpm. In addition, two-winding motors can be wound to provide two speeds on each winding; this makes it possible for the motor to

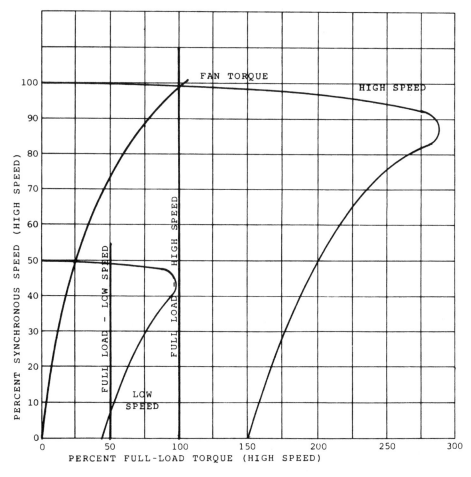

FIGURE 1.7 Speed-torque curves for a variable-torque, one-winding, two-speed motor.

operate at four speeds, for example, 3600/1800 rpm on one winding and 1200/600 rpm on the other winding.

Multispeed motors are available with the following torque characteristics.

Variable Torque.   The variable-torque multispeed motor has a torque output that varies directly with the speed, and hence the

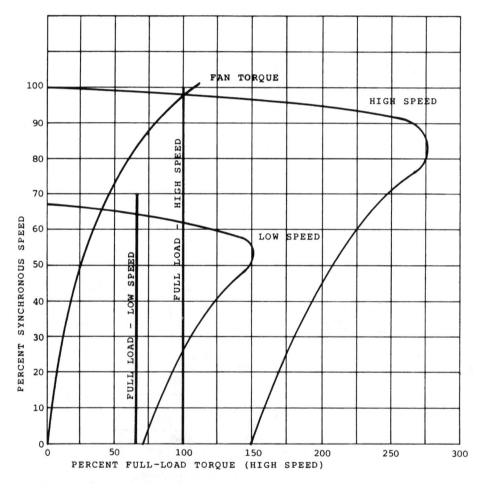

FIGURE 1.8   Speed-torque curves for a multispeed variable-torque motor with two windings, two speeds, and a four-pole to six-pole ratio.

horsepower output varies with the square of the speed. This motor is commonly used with fans, blowers, and centrifugal pumps to control the output of the driven device. Figure 1.7 shows typical speed-torque curves for this type of motor. Superimposed on the motor speed-torque curve is the speed-torque curve for a typical fan where the input horsepower to the fan varies as the cube of the fan speed. Another popular drive for fans is a two-winding two-speed

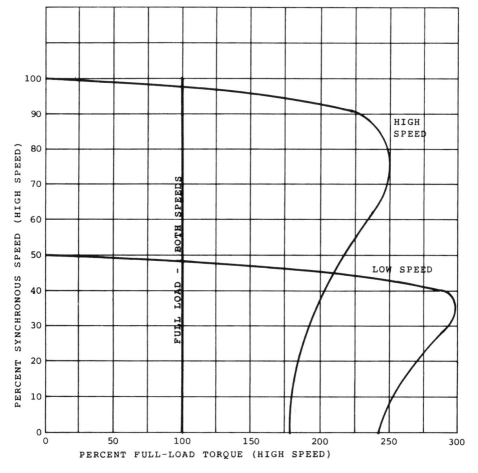

FIGURE 1.9 Speed-torque curves for a constant-torque, one-winding, two-speed motor.

motor, such as 1800 rpm at high speed and 1200 rpm at low speed. Figure 1.8 shows the typical motor speed-torque curve for the two-winding variable-torque motor with a fan speed-torque curve superimposed.

Constant Torque.   The constant-torque multispeed motor has a torque output that is the same at all speeds, and hence the horsepower

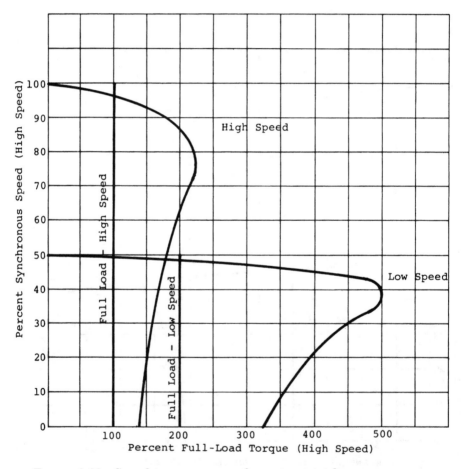

FIGURE 1.10 Speed-torque curves for a constant-horsepower, one-winding two-speed motor.

output varies directly with the speed. This motor can be used with friction-type loads such as those found on conveyors to control the conveyor speed. Figure 1.9 shows typical speed-torque curves.

Constant Horsepower. The constant-horsepower multispeed motor has the same horsepower output at all speeds. This type of motor is used for machine tool applications that require higher torques at lower speeds. Figure 1.10 shows typical speed-torque curves.

1.2 SINGLE-PHASE INDUCTION MOTORS

There are many types of single-phase electric motors. In this section, the discussion will be limited to those types most common to integral-horsepower motor ratings of 1 hp and higher.

In industrial applications, three-phase induction motors should be used wherever possible. In general, three-phase electric motors have higher efficiency and power factors and are more reliable since they do not have starting switches or capacitors.

In those instances in which three-phase electric motors are not available or cannot be used because of the power supply, the following types of single-phase motors are recommended for industrial and commercial applications: (1) capacitor-start motor, (2) two-value capacitor motor, and (3) permanent split capacitor motor.

A brief comparison of single-phase and three-phase induction motor characteristics will provide a better understanding of how single-phase motors perform:

1. Three-phase motors have locked torque because there is a revolving field in the air gap at standstill. A single-phase motor has no revolving field at standstill and therefore develops no locked-rotor torque. An auxiliary winding is necessary to produce the rotating field required for starting. In an integral-horsepower single-phase motor, this is part of an RLC network.
2. The rotor current and rotor losses are insignificant at no load in a three-phase motor. Single-phase motors have appreciable rotor current and rotor losses at no load.

3. For a given breakdown torque, the single-phase motor requires considerably more flux and more active material than the equivalent three-phase motor.
4. A comparison of the losses between single-phase and three-phase motors is shown in Fig. 1.11. Note the significantly higher losses in the single-phase motor.

The general characteristics of these types of single-phase induction motors are as follows.

1.2.1 Capacitor-Start Motors

A capacitor-start motor is a single-phase induction motor with a main winding arranged for direct connection to the power source and an auxiliary winding connected in series with a capacitor and starting switch for disconnecting the auxiliary winding from the power source after starting. Figure 1.12 is a schematic diagram of a capacitor-start motor. The type of starting switch most commonly

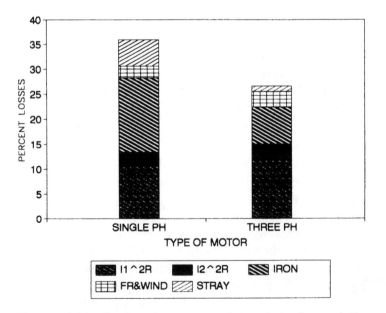

FIGURE 1.11 Percent loss comparison of single- and three-phase motors.

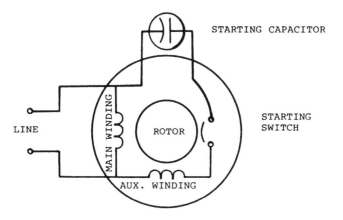

FIGURE 1.12 Capacitor-start single-phase motor.

used is a centrifugally actuated switch built into the motor. Figure 1.13 illustrates an industrial-quality drip-proof single-phase capacitor-start motor; note the centrifugally actuated switch mechanism.

However, other types of devices such as current-sensitive and voltage-sensitive relays are also used as starting switches. More recently, solid-state switches have been developed and used to a

FIGURE 1.13 Capacitor-start single-phase motor. (Courtesy Magnetek, St. Louis, MO.)

limited extent. The solid-state switch will be the switch of the future as it is refined and costs are reduced.

All the switches are set to stay closed and maintain the auxiliary winding circuit in operation until the motor starts and accelerates to approximately 80% of full-load speed. At that speed, the switch opens, disconnecting the auxiliary winding circuit from the power source.

The motor then runs on the main winding as an induction motor. The typical speed-torque characteristics for a capacitor-start motor are shown in Fig. 1.14. Note the change in motor torques at the transition point at which the starting switch operates.

The typical performance data for integral-horsepower, 1800-rpm, capacitor-start, induction-run motors are shown in Table 1.6. There will be a substantially wider variation in the values of locked-rotor torque, breakdown torque, and pull-up torque for these single-phase motors than for comparable three-phase motors, and the same variation also exists for efficiency and the power factor (PF). Note that pull-up torque is a factor in single-phase motors to ensure starting with high-inertia or hard-to-start loads. Therefore, it is important to know the characteristics of the specific capacitor-start motor to make certain it is suitable for the application.

1.2.2 Two-Value Capacitor Motors

A two-value capacitor motor is a capacitor motor with different values of capacitance for starting and running. Very often, this type of motor is referred to as a capacitor-start, capacitor-run motor.

The change in the value of capacitance from starting to running conditions is automatic by means of a starting switch, which is the same as that used for the capacitor-start motors. Two capacitors are provided, a high value of capacitance for starting conditions and a lower value for running conditions. The starting capacitor is usually an electrolytic type, which provides high capacitance per unit volume. The running capacitor is usually a metallized polypropylene unit rated for continuous operation. Figure 1.15 shows one method of mounting both capacitors on the motor.

The schematic diagram for a two-value capacitor motor is shown in Fig. 1.16. As shown, at starting, both the starting and

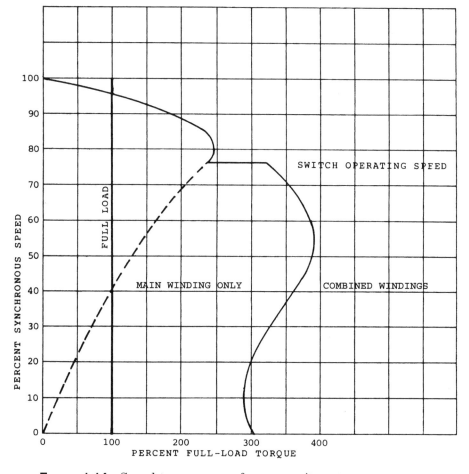

FIGURE 1.14 Speed-torque curve for a capacitor-start motor.

running capacitors are connected in series with the auxiliary winding. When the starting switch opens, it disconnects the starting capacitor from the auxiliary winding circuit but leaves the running capacitor in series with the auxiliary winding connected to the power source. Thus, both the main and auxiliary windings are energized when the motor is running and contribute to the motor output. A

TABLE 1.6   Typical Performance of Capacitor-Start Motors[a]

| | Full-load performance | | | | | Torque, lb-ft | | |
hp	rpm	A	Eff.	PF	Torque	Locked	Breakdown	Pull-up
1	1725	7.5	71	70	3.0	9.9	7.5	7.6
2	1750	12.5	72	72	6.0	17.5	14.7	11.5
3	1750	17.0	74	79	9.0	23.0	21.0	18.5
5	1745	27.3	78	77	15.0	46.0	32.0	35.0

[a] Four-pole, 230-V, single-phase motors.
Source: Courtesy Magnetek, St. Louis, MO.

typical speed-torque curve for a two-valve capacitor motor is shown in Fig. 1.17.

For a given capacitor-start motor, the effect of adding a running capacitor in the auxiliary winding circuit is as follows:

Increased breakdown torque: 5–30%
Increased lock-rotor torque: 5–10%
Improved full-load eciency: 2–7 points

FIGURE 1.15   Two-value capacitor, single-phase motor. (Courtesy Magnetek, St. Louis, MO.)

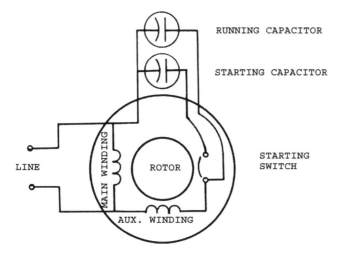

FIGURE 1.16 Two-value capacitor, single-phase motor.

Improved full-load power factor: 10–20 points
Reduced full-load running current
Reduced magnetic noise
Cooler running

The addition of a running capacitor to a single-phase motor with properly designed windings permits the running performance to approach the performance of a three-phase motor. The typical performance of integral-horsepower, two-value capacitor motors is shown in Table 1.7. Comparison of this performance with the performance shown in Table 1.6 for capacitor-start motors shows the improvement in both efficiency and the power factor.

The optimum performance that can be achieved in a two-value capacitor, single-phase motor is a function of the economic factors as well as the technical considerations in the design of the motor. To illustrate this, Table 1.8 shows the performance of a single-phase motor with the design optimized for various values of running capacitance. The base for the performance comparison is a capacitor-start, induction-run motor with no running capacitor. Table 1.9 shows that performance improves with increasing values of running

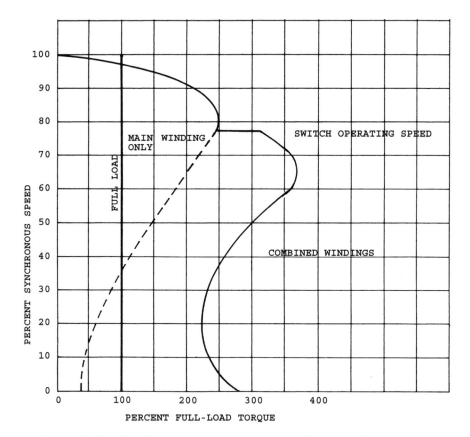

FIGURE 1.17 Speed-torque curve for a two-value capacitor motor.

capacitance and that the motor costs increase as the value of running capacitance is increased. The payback period in years was calculated on the basis of 4000 hr/yr of operation and an electric power cost of 6¢/kWh. Note that the major improvement in motor performance is made in the initial change from a capacitor-start to a two-value capacitor motor with a relatively low value of running capacitance. This initial design change also shows the shortest payback period.

The determination of the optimum two-value capacitor motor for a specific application requires a comparison of the motor costs and the energy consumptions of all such available motors. It is

TABLE 1.7 Typical Performance of Two-Value Capacitor Motors[a]

	Full-load performance					Torque, lb/ft		
hp	rpm	A	Eff.	PF	Torque	Locked	Breakdown	Pull-up
3	1760	14.0	78	90	9.0	25	23	22
5	1760	25.0	82	80	15.0	46	35	32
7.5	1750	32.0	86	88	22.5	45	56	45
10	1750	38.0	86	96	30.0	56	72	56

[a] Four-pole, 230-V, single-phase motors.
Source: Courtesy Magnetek, St. Louis, MO.

recommended that this comparison be made by a life-cycle cost method or the net present worth method (outlined in Chapter 7).

The efficiency improvement and energy savings of a specific product line of pool pump motors when the design was changed from capacitor-start motors to two-value capacitor motors are illustrated by Table 1.9 and Figs. 1.18 and 1.19. Based on the same operating criterion used above, i.e., 4000-hr/yr operation at power costs of 6¢/kWh, the payback period for these motors was 8–20 months.

TABLE 1.8 Performance Comparison of Capacitor-Start and Two-Value Capacitor Motors

	Type of motor				
	Capacitor start	Two-value capacitor			
Running capacitor, MFD	0	7.5	15	30	65
Full-load efficiency	70	78	79	81	83
Full-load PF	79	94	97	99[a]	99[a]
Input watts reduction, %	0	10.1	11.5	13.3	15
Cost, %	100	130	140	151	196
Approximate payback period	—	1.3	1.6	1.8	2.9

[a] Leading power factor.

TABLE 1.9 Efficiency Comparison: Standard and Energy-Efficient 3600-rpm, Single-Phase Pool Motors

hp	Standard efficient motors	Energy-efficient motors
0.75	0.677	0.76
1.00	0.709	0.788
1.50	0.749	0.827
2.00	0.759	0.85
3.00	0.809	0.869

Source: Courtesy Magnetek, St. Louis, MO.

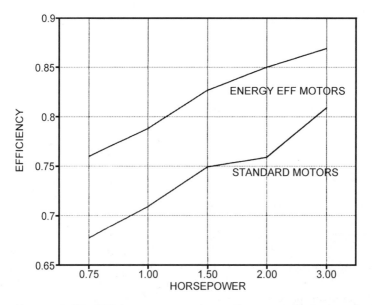

FIGURE 1.18 Efficiency comparison of energy-efficient and standard pool pump single-phase motors. (Courtesy Magnetek, St. Louis, MO.)

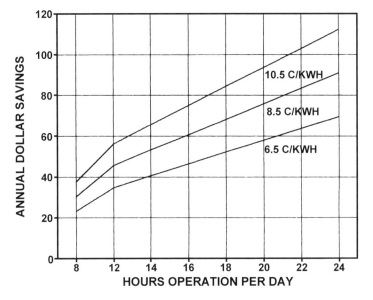

FIGURE 1.19 Annual savings for a 1-hp energy-efficient pool motor operating 365 days/yr. (Courtesy Magnetek, St. Louis, MO.)

1.2.3 Permanent Split Capacitor Motors

The permanent split capacitor motors, a single-phase induction motor, is defined as a capacitor motor with the same value of capacitance used for both starting and running operations. This type of motor is also referred to as a single-value capacitor motor. The application of this type of single-phase motor is normally limited to the direct drive of such loads as those of fans, blowers, or pumps that do not require normal or high starting torques. Consequently, the major application of the permanent split capacitor motor has been to direct-driven fans and blowers. These motors are not suitable for belt-driven applications and are generally limited to the lower horsepower ratings.

The schematic diagram for a permanent split capacitor motor is shown in Fig. 1.20. Note the absence of any starting switch. This type of motor is essentially the same as a two-value capacitor motor

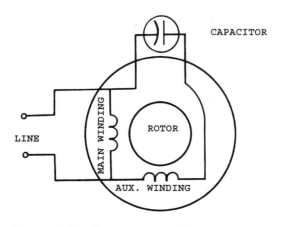

Figure 1.20 Permanent split capacitor single-phase motor.

operating on the running connection and will have approximately the same torque characteristics. Since only the running capacitor (which is of relative low value) is connected in series with the auxiliary winding on starting, the starting torque is greatly reduced. The starting torque is only 20–30% of full-load torque. A typical speed-torque curve for a permanent split capacitor motor is shown in Fig. 1.21. The running performance of this type of motor in terms of efficiency and power factor is the same as a two-value capacitor motor. However, because of its low starting torque, its successful application requires close coordination between the motor manufacturer and the manufacturer of the driven equipment.

A special version of the capacitor motor is used for multiple-speed fan drives. This type of capacitor motor usually has a tapped main winding and a high-resistance rotor. The high-resistance rotor is used to improve stable speed operation and to increase the starting torque. There are a number of versions and methods of winding motors. The most common design is the two-speed motor, which has three windings: the main, intermediate, and auxiliary windings. For 230-V power service, a common connection of the windings is called the T connection. Schematic diagrams for two-speed T-connected motors are shown in Figs. 1.22 and 1.23. For high-speed operation,

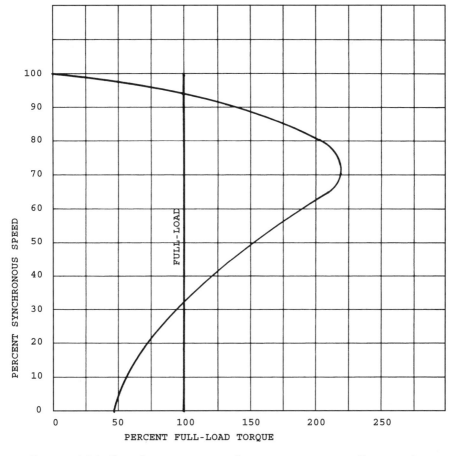

FIGURE 1.21 Speed-torque curve for a permanent split capacitor motor.

the intermediate winding is not connected in the circuit as shown in Fig. 1.23, and line voltage is applied to the main winding and to the auxiliary winding and capacitor in series. For low-speed operation, the intermediate winding is connected in series with the main winding and with the auxiliary circuit as shown in Fig. 1.23. This connection reduces the voltage applied across both the main wind-

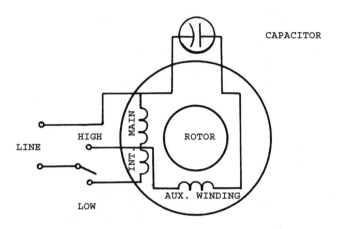

FIGURE 1.22 Permanent split capacitor single-phase motor with a T-type connection and two-speed operation.

ing and the auxiliary circuit, thus reducing the torque the motor will develop and hence the motor speed to match the load requirements. The amount of speed reduction is a function of the turns ratio between the main and intermediate windings and the speed-torque characteristics of the driven load. It should be recognized that, with this type of motor, the speed change is obtained by letting the motor

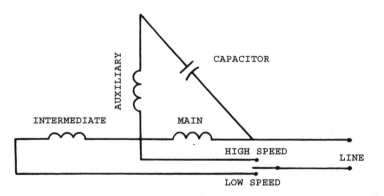

FIGURE 1.23 Permanent split capacitor single-phase motor with a T-type connection and a winding arrangement.

speed slip down to the required low speed; it is not a multispeed motor with more than one synchronous speed.

An example of the speed-torque curves for a tapped-winding capacitor motor is shown in Fig. 1.24. The load curve of a typical fan load is superimposed on the motor speed-torque curves to show the speed reduction obtained on the low-speed connection.

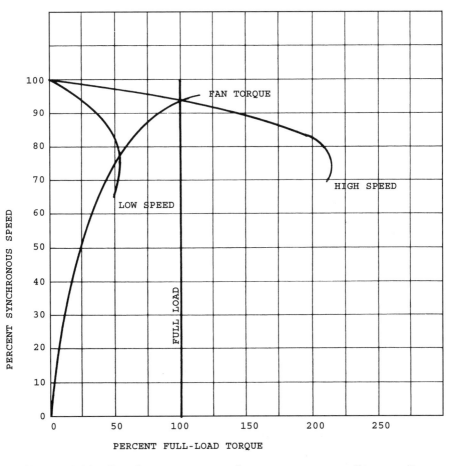

FIGURE 1.24 Speed-torque curves for a permanent split capacitor single-phase motor with a tapped winding.

2

Energy-Efficient Motors

2.1 STANDARD MOTOR EFFICIENCY

During the period from 1960 to 1975, electric motors, particularly those in the 1- to 250-hp range, were designed for minimum first cost. The amount of active material, i.e., lamination steel, copper or aluminum or magnet wire, and rotor aluminum, was selected as the minimum levels required to meet the performance requirements of the motor. Efficiency was maintained at levels high enough to meet the temperature rise requirements of the particular motor. As a consequence, depending on the type of enclosure and ventilation system, a wide range in efficiencies exists for standard NEMA design B polyphase motors. Table 2.1 is an indication of the range of the nominal electric motor efficiencies at rated horsepower. These data are also presented in Fig. 2.1. The data are based on information published by the major electric motor manufacturers. However, the meaning or interpretation of data published prior to the NEMA adoption of the definition of nominal efficiency is not always clear. In 1977, NEMA recommended a procedure for marking the three-

TABLE 2.1 Full-Load Efficiencies of NEMA Design B Standard Three-Phase Induction Motors

hp	Nominal efficiency range	Average nominal efficiency
1	68–78	73
1.5	68–80	75
2	72–81	77
3	74–83	80
5	78–85	82
7.5	80–87	84
10	81–88	85
15	83–89	86
20	84–89	87.5
25	85–90	88
30	86–90.5	88.5
40	87–91.5	89.5
50	88–92	90
60	88.5–92	90.5
75	89.5–92.5	91
100	90–93	91.5
125	90.5–93	92
150	91–93.5	92.5
200	91.5–94	93
250	91.5–94.5	93.5

phase motors with a NEMA nominal efficiency. This efficiency represents the average efficiency for a large population of motors of the same design. In addition, a minimum efficiency was established for each level of nominal efficiency.

The minimum efficiency is the lowest level of efficiency to be expected when a motor is marked with the nominal efficiency in accordance with the NEMA standard. This method of identifying the motor efficiency takes into account variations in materials, manufacturing processes, and test results in motor-to-motor efficiency variations for a given motor design. The nominal efficiency represents a value that should be used to compute the energy consump-

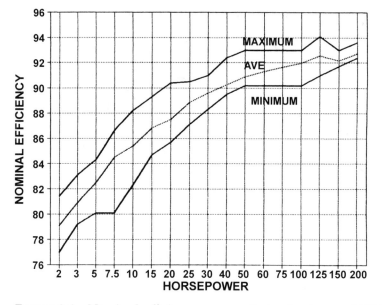

FIGURE 2.1 Nominal efficiency range of standard open NEMA design B 1800-rpm polyphase induction motors.

tion of a motor or group of motors. Table 2.1 shows a wide range in efficiency for individual motors and, consequently, a range in the electric motor losses and electric power input. For example, a standard 10-hp electric motor may have an efficiency range of 81–88%.

At 81% efficiency,

$$\text{Input electric power} = \frac{10 \times 746}{0.81} = 9210 \text{ W}$$

$$\text{Motor losses} = 9210 - 7460 = 1750 \text{ W}$$

At 88% efficiency,

$$\text{Input electric power} = \frac{10 \times 746}{0.88} = 8477 \text{ W}$$

$$\text{Motor losses} = 8477 - 7460 = 1017 \text{ W}$$

Therefore, for the same output the input can range from 8477 to 9210 W, or an increase in energy consumption and power costs of 8%, to operate the less efficient motor.

2.2 WHY MORE EFFICIENT MOTORS?

The escalation in the cost of electric power that began in 1972 made it increasingly expensive to use inefficient electric motors. From 1972 through 1979, electric power rates increased at an average annual rate of 11.5%/yr. From 1979 to the present, the electric power rates have continued to increase at an average annual rate of 6%/yr. The annual electric power cost to operate a 10-hp motor 4000 hr/yr increased from $850 in 1972 to $1950 in 1980 and to over $2500 by 1989. By 1974, electric motor manufacturers were looking for methods to improve three-phase induction motor efficiencies to values above those shown for standard NEMA design B motors in Table 2.1.

2.3 WHAT IS EFFICIENCY?

Electric motor efficiency is the measure of the ability of an electric motor to convert electrical energy to mechanical energy; i.e., kilowatts of electric power are supplied to the motor at its electrical terminals, and the horsepower of mechanical energy is taken out of the motor at the rotating shaft. Therefore, the only power absorbed by the electric motor is the losses incurred in making the conversion from electrical to mechanical energy. Thus, the motor efficiency can be expressed as

$$\text{Efficiency} = \frac{\text{mechanical energy out}}{\text{electrical energy in}} \times 100\%$$

but

$$\text{Mechanical energy out} = \text{electrical energy in} - \text{motor losses}$$

or

$$\text{Electrical energy in} = \text{mechanical energy out} + \text{motor losses}$$

Therefore, to reduce the electric power consumption for a given mechanical energy out, the motor losses must be reduced and the electric motor efficiency increased.

To accomplish this, it is necessary to understand the types of losses that occur in an electric motor. These losses consist of the following.

2.3.1 Power Losses

The power losses (I^2R in the motor windings) consist of two losses: the stator power losses I^2R and the rotor power losses I^2R. The stator power loss is a function of the current flowing in the stator winding and the stator winding resistance—hence the term I^2R loss:

$$\text{Stator current } I_1 = \frac{\text{input watts}}{\text{voltage} \times \sqrt{3} \text{ PF}}$$

When improving the motor performance, it is important to recognize the interdependent relationship of the efficiency and the power factor. Rewrite the preceding equation and solve for the power factor:

$$\text{PF} = \frac{\text{output hp} \times 746}{\text{voltage} \times \sqrt{3} \text{ efficiency} \times I_1}$$

Therefore, if the efficiency is increased, the power factor will tend to decrease. For the power factor to remain constant, the stator current I_1 must decrease in proportion to the increase in efficiency. To increase the power factor, the stator current must be decreased more than the efficiency is increased. From a design standpoint, this is difficult to accomplish and still maintain other performance requirements such as breakdown torque. However,

$$\text{Input watts} = \frac{\text{output hp} \times 746}{\text{efficiency}}$$

or

$$I_1 = \frac{\text{output hp} \times 746}{\text{voltage} \times \sqrt{3} \text{ PF} \times \text{efficiency}}$$

Therefore, the stator losses are inversely proportional to the square of the efficiency and the power factor. In addition, the stator loss is a function of the stator winding resistance. For a given configuration, the winding resistance R is inversely proportional to the pounds of

magnet wire or conductors in the stator winding. The more conductor material in the stator winding, the lower the losses.

The rotor power loss is generally expressed as the slip loss:

$$\text{Rotor loss} = \frac{(\text{hp output} \times 746 + \text{FW})S}{1 - S}$$

$$S = \frac{N_s - N}{N_s} = \text{slip}$$

where

N = output speed, rpm
N_s = synchronous speed, rpm
FW = friction and windage loss

The rotor slip can be reduced by increasing the amount of
conductor material in the rotor or increasing the total flux across the
air gap into the rotor. The extent of these changes is limited by the
minimum starting (or locked-rotor) torque required, the maximum
locked-rotor current, and the minimum power factor required.

2.3.2 Magnetic Core Losses

Magnetic core losses consist of the eddy current and hysteresis
losses, including the surface losses, in the magnetic structure of the
motor. A number of factors influence these losses:

1. The flux density in the magnetic structure is a major
    factor in determining these magnetic losses. The core
    loss can be decreased by increasing the length of the
    magnetic structure and, as a consequence, decreasing
    the flux density in the core. This will decrease the
    magnetic loss per unit of weight but, since the total
    weight will increase, the improvement in losses will
    not be proportional to the unit loss reduction. The de-
    crease in magnetic loading in the motor also decreases
    the magnetizing current and thus influences the
    power factor.
2. The magnetic core loss can also be reduced by using
    thinner laminations in the magnetic structure. Typi-

cally, many standard motors use 24-gauge (0.025-in. thick) laminations. By using thinner laminations, such as 26-gauge (0.0185-in. thick) or 29-gauge (0.014-in. thick), the magnetic core loss can be reduced. The reduction in the magnetic core loss by the use of thinner laminations ranges from 10 to 25%, depending on the method of processing the lamination steel and the method of assembling the magnetic core.

3. There has been considerable progress made by the steel companies to obtain lower magnetic losses in both silicon and cold-rolled (low-silicon) grades of electrical steel. The magnetic core loss (Epstein loss) can be reduced by using silicon grades of electrical steel or the improved grades of cold-rolled electrical steel. The type of steel used by the motor manufacturer depends on his process capability. The cold-rolled electrical steel requires a proper anneal after punching to develop its electrical properties, whereas the silicon grades of electrical steel are available as fully processed material. Tables 2.2a and 2.2b illustrate some of the silicon and cold-rolled electrical steels available and the influence of grade and thickness on the Epstein loss and permeability.

However, because of variables in the processing of the lamination steel into finished motor cores, the reduction in core loss in watts per pound equivalent to the Epstein data on flat strips of the lamination steel is seldom achieved. Magnetic core loss reductions on the order of 15–40% can be achieved by the use of thinner-gauge silicon-grade electrical steels. A disadvantage of the higher-silicon lamination steel is that, at high inductions, the permeability may be lower, thus increasing the magnetizing current required. This will tend to decrease the motor power factor.

2.3.3 Friction and Windage Losses

Friction and windage losses are caused by the friction in the bearings of the motor and the windage loss of the ventilation fan and

TABLE 2.2a Typical 50/50 as Sheared Epstein Data for Silicon-Grade Electrical Steel

Electrical steel grade	Standard gage	Nominal thickness, in.	Maximum epstein loss at 15 kg, 60 Hz	Typical permeability at 15 kg, 60 Hz
M-47	24	0.025	3.60	1800
	26	0.0185	3.05	1800
M-45	24	0.025	3.20	1700
	26	0.0185	2.80	1700
M-43	24	0.025	2.70	1500
	26	0.0185	2.30	1500
	29	0.014	2.00	1500
M-36	26	0.0185	2.05	1400
	29	0.014	1.90	1400
M-27	29	0.014	1.80	1200

Note: The Epstein core loss is for fully processed steel; lower losses can be attained with semiprocessed steel and a quality anneal.
Source: Courtesy Armco Advanced Materials Co., Butler, PA.

other rotating elements of the motor. The friction losses in the bearings are a function of bearing size, speed, type of bearing, load, and lubrication used. This loss is relatively fixed for a given design and, since it is a small percentage of the total motor losses, design changes to reduce this loss do not significantly affect the motor efficiency. Most of the windage losses are associated with the ventilation fans and the amount of ventilation required to remove the heat generated by other losses in the motor, such as the winding power losses I^2R, magnetic core loss, and stray load loss. As the heat-producing losses are reduced, it is possible to reduce the ventilation required to remove those losses, and thus the windage loss can be reduced. This applies primarily to totally enclosed fan-cooled motors with external ventilation fans. One of the important by-products of decreasing the windage loss is a lower noise level created by the motor.

TABLE 2.2b Typical Epstein Data Inland Steel Nonsilicon Cold-Rolled Electrical Steel

Inland type	Thickness, in.	Epstein loss, W/lb at 15 kg	Permeability at 15 kg (min.)
Rephosphorized	0.029	4.6	2000
	0.025	3.85	2000
	0.022	3.5	2000
Interlocking	0.029	4.2	2000
	0.025	3.75	2000
	0.022	3.2	2000
2.5/2000	0.025	3.3	2000
	0.022	2.9	2000
	0.018	2.5	2000
2.25/2000	0.025	3.1	2000
	0.022	2.7	2000
	0.018	2.3	2000
2/2000	0.022	2.3	2000
	0.018	2.0	2000

Note: The Epstein values are typical for semiprocessed steel annealed after punching.
Source: Courtesy Inland Steel Flat Products Co., Chicago, IL.

2.3.4 Stray Load Losses

Stray load losses are residual losses in the motor that are difficult to determine by direct measurement or calculation. These losses are load related and are generally assumed to vary as the square of the output torque. The nature of this loss is very complex. It is a function of many of the elements of the design and the processing of the motor. Some of the elements that influence this loss are the stator winding design, the ratio of air gap length to rotor slot openings, the ratio of the number of rotor slots to stator slots, the air gap flux density, the condition of the stator air gap surface, the condition of the rotor air gap surface, and the bonding or welding of the rotor conductor bars to rotor lamination. By careful design, some of the elements that contribute to the stray loss can be minimized. Those stray losses that relate to processing, such as surface conditions, can be minimized by careful manufacturing process control. Because of the large number of variables that contribute to the stray loss, it is the most difficult loss in the motor to control.

2.3.5 Summary of Loss Distribution

Within a limited range, the various motor losses discussed are independent of each other. However, in trying to make major improvements in efficiency, one finds that the various losses are very dependent. The final motor design is a balance among several losses to obtain a high efficiency and still meet other performance criteria, including locked-rotor torque, locked-rotor amperes, breakdown torque, and the power factor.

The distribution of electric motor losses at the rated load is shown in Table 2.3 for several horsepower ratings. It is important for the motor designer to understand this loss distribution in order to make design changes to improve motor efficiency. In a very general sense, the average loss distribution for standard NEMA design B motors can be summarized as follows:

Motor component loss	Total loss, %
Stator power loss I^2R	37
Rotor power loss I_2^2R	18
Magnetic core loss	20
Friction and windage	9
Stray load loss	16

This loss distribution indicates the significance of design changes to increase the electric motor efficiency. However, as the motor efficiency and the horsepower increase, the level of difficulty

TABLE 2.3 Typical Loss Distribution of Standard NEMA Design B
Drip-Proof Motors

| | | 1 hp | | | 5 hp | |
Loss distribution	Watts	% Loss	PU loss	Watts	% Loss	PU loss
Stator power loss I_1^2R	120	43	0.16	305	40	0.08
Rotor power loss I_2^2R	35	13	0.05	150	20	0.04
Magnetic core loss	76	28	0.10	225	29	0.06
Friction and windage loss	24	9	0.03	30	4	0.01
Stray load loss	19	7	0.03	51	7	0.01
Total losses	274	100	0.37	761	100	0.20
Output, W	746			3730		
Input, W	1020			4491		
Efficiency, %	73			83		

Notes: Polyphase four-pole motor, 1750 rpm. % loss = percent of total losses.
PU loss = loss/(hp × 746).

in improving the electric motor efficiency increases. Consider the
stator and rotor power losses only. To improve the motor full-load
efficiency, one efficiency point requires an increasing reduction in
these power losses as the motor efficiency increases:

hp	Original efficiency, %	Increased efficiency, %	Decrease in power losses required, %
1	73.0	74.0	8
5	83.0	84.0	11
25	89.0	90.0	16
50	90.5	91.5	19
100	91.5	92.5	28
200	93.0	94.0	38

These loss reductions can be achieved by increasing the amount
of material, i.e., magnet wire in the stator winding and aluminum

25 hp			50 hp			100 hp			200 hp		
Watts	% Loss	PU loss	Watts	% Loss	PU loss	Watts	% Loss	PU loss	Watts	% Loss	PU loss
953	42	0.05	1,540	38	0.04	1,955	28	0.026	3,425	30	0.023
479	21	0.03	860	22	0.02	1,177	18	0.016	1,850	16	0.012
351	15	0.02	765	20	0.02	906	13	0.012	1,650	15	0.011
168	7	0.01	300	8	0.01	992	14	0.013	1,072	10	0.007
345	15	0.02	452	12	0.01	1,900	27	0.025	3,235	29	0.022
2,296	100	0.13	3,917	100	0.10	6,930	100	0.092	11,232	100	0.075
18,560			37,300			74,600			149,200		
20,946			41,217			81,530			160,432		
89			90.5			91.5			93.0		

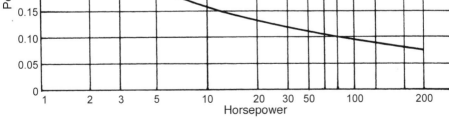

FIGURE 2.2 Per unit losses for standard design B four-pole motors.

conductors in the rotor or squirrel-cage winding. However, a loss deduction of only 5–15% can be achieved in these power losses without making other design modifications. These modifications can include a new lamination design to increase the amount of magnet wire and aluminum rotor conductors that can be used, combined with the use of lower-loss electrical-grade lamination steel in the magnetic structure and the use of a longer magnetic structure. The level of difficulty and, consequently, the cost of improving the electric motor efficiency increases as the horsepower rating increases. This is illustrated in Fig. 2.2, which shows the decrease in per unit losses as the horsepower rating increases, thus requiring a larger per unit loss reduction at the higher horsepower ratings for the same efficiency improvement.

2.4   WHAT IS AN ENERGY-EFFICIENT MOTOR?

Until recently, there was no single definition of an energy-efficient motor. Similarly, there were no efficiency standards for standard NEMA design B polyphase induction motors. As discussed earlier, standard motors were designed with efficiencies high enough to achieve the allowable temperature rise for the rating. Therefore, for a given horsepower rating, there is a considerable variation in efficiency. This is illustrated in Fig. 2.1 for the horsepower range of 1–200 hp.

In 1974, one electric motor manufacturer examined the trend of increasing energy costs and the costs of improving electric motor efficiencies. The cost/benefit ratio at that time justified the development of a line of energy-efficient motors with losses approximately 25% lower than the average NEMA design B motors. This has resulted in a continuing industry effort to decrease the watt losses of induction motors. Figure 2.3 shows a comparison between the full-load watt losses for standard four-pole, 1800-rpm NEMA design B induction motors, the first-generation energy-efficient motors with a 25% reduction in watt losses, and the current energy-

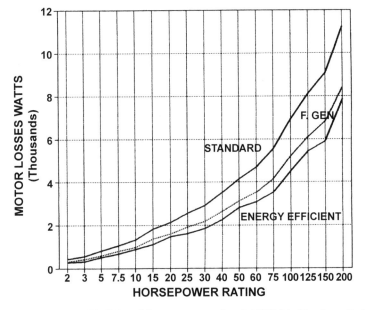

FIGURE 2.3 Full-load losses, standard NEMA Design B 1800-rpm motors versus first-generation energy-efficient motors (25% loss reduction) and current energy-efficient motors.

efficient motors. The watt loss reduction for the current energy-efficient four-pole, 1800-rpm motors ranges from 25 to 43%, with an average watt loss reduction of 35%. Figures 2.4a and 2.4b illustrate the nominal efficiencies of the current energy-efficient (E.E.) motors, the first-generation energy-efficient motors (25% loss reduction), and current standard NEMA design B four-pole, 1800-rpm motors.

Subsequent to the development of this first line of energy-efficient motors, all major electric motor manufacturers have followed suit. Since, as previously discussed, there was no standard for the efficiency of motors, the energy-efficient motors of the various manufacturers can generally be identified by their trade names. In addition, these products are supported by appropriate published

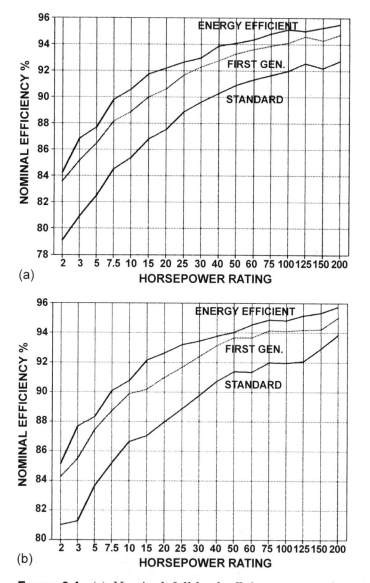

FIGURE 2.4 (a) Nominal full-load efficiency comparison 1800-rpm open induction motors. (b) Nominal full-load efficiency comparison 1800-rpm TEFC induction motors.

data. Following are examples of these trade names and their manufacturers:

E-Plus®, E + III®	Magnetek
Energy Saver™	General Electric
XE Energy Efficient	Reliance Electric Co.
Super E	Baldor Electric Co.
Spartan High Efficiency	Magnetek/Louis Allis
Corro-Duty Premium Efficiency	U.S. Electrical Motors
Premium Efficiency	Division of Emerson Electric Co. Seimens *misspelled*
Premium Efficiency	Toshiba/Houston Intl.

A survey of the published data available from the manufacturers of energy-efficient motors is summarized in Table 2.4 and Fig. 2.5. These data show the nominal average efficiency as well as the range of nominal efficiencies expected. The efficiencies are shown as nominal efficiencies as defined in NEMA Standards Publication MG1. When these efficiency data are compared to the standard motor efficiency data shown in Fig. 2.1, the range in efficiency for a given horsepower is considerably less; in other words, energy-efficient motors tend to be more uniform than standard motors.

When the average nominal efficiency for industry energy-efficient motors shown in Tables 2.4a and 2.4b is compared to the data shown in Fig. 2.4 for standard motors, the industry average is consistently higher. When the average efficiency for standard motors in Fig. 2.1 is compared to the average efficiency for current energy-efficient motors in Figs. 2.5a and 2.5b the average loss reduction is 35%, thus indicating a continuing trend to higher-efficiency motors. These improvements in efficiency, or loss reductions, are generally achieved by increasing the amount of active material used in the motors and by the use of lower-loss magnetic steel. Figure 2.6 shows this comparison of a standard motor and an energy-efficient motor for a particular horsepower rating. In addition to increasing the motor efficiency, there are other user benefits in the application of energy-efficient motors, which will be discussed in more detail in

TABLE 2.4a Full-Load Nominal Efficiencies of Three-Phase Four-Pole Energy-Efficient Open Motors[a]

	Nominal efficiency range		Average nominal efficiency
hp	Min.	Max.	
1	82.5	85.5	83.5
2	82.5	86.0	84.3
3	82.5	89.5	87.0
5	85.5	89.5	87.9
7.5	87.5	91.7	90.0
10	89.5	91.7	90.6
15	90.2	93.0	91.6
20	90.2	93.6	92.1
25	91.0	94.1	92.6
30	91.7	94.1	93.1
40	93.6	94.5	93.9
50	93.6	94.5	94.1
60	93.6	95.4	94.4
75	94.1	95.4	94.8
100	94.1	96.2	95.1
125	94.1	95.4	95.1
150	94.5	96.2	95.3
200	95.0	96.2	95.5

[a] Based on available published data.

Chapter 5. This trend will probably continue as the cost of power and the demand for higher-efficiency motors continue to increase. Figure 2.7 shows the trend in the loss reduction and efficiency improvement of a 50-hp polyphase induction motor. Other induction motors from 1 to over 200 hp have followed a similar trend.

2.5   EFFICIENCY DETERMINATION

Efficiency is defined as the ratio of the output power to the input power to the motor expressed in percent; thus,

$$\text{Percent efficiency} = \frac{W_{out} \times 100}{W_{in}}$$

TABLE 2.4b Full-Load Nominal Efficiencies of Three-Phase Four-Pole Energy-Efficient TEFC Motors

	Nominal efficiency range		Average nominal efficiency
hp	Min.	Max.	
1	77.0	85.5	82.5
2	82.5	86.5	84.2
3	82.5	89.5	87.8
5	85.5	90.2	88.4
7.5	87.5	91.0	90.3
10	89.5	91.7	90.6
15	91.0	92.5	92.0
20	91.9	93.0	92.6
25	92.3	94.1	93.1
30	92.6	94.1	93.4
40	93.1	94.5	93.8
50	93.2	95.0	94.0
60	93.6	95.0	94.4
75	93.8	95.4	94.7
100	94.1	95.4	94.9
125	94.2	95.8	95.2
150	94.2	96.2	95.4
200	95.2	96.2	95.7

[a]Based on available published data.

It may also be expressed as

$$\text{Percent efficiency} = \frac{W_{out} \times 100}{W_{out} + W_{loss}}$$

where

W_{out} = output power, W
W_{in} = input power, W
W_{loss} = motor losses, W

The total motor losses include the following losses:

$$W_{loss} = W_s + W_r + W_c + W_f + W_{sl}$$

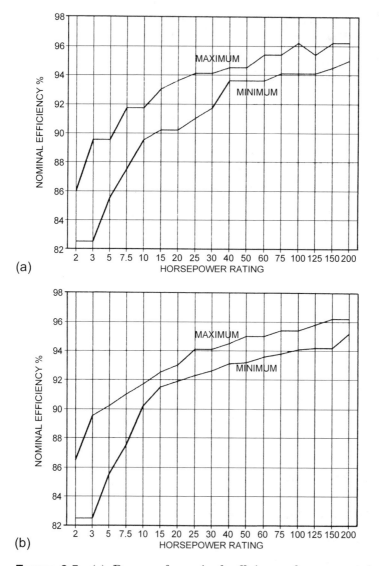

FIGURE 2.5 (a) Range of nominal efficiency for current industry energy-efficient open 1800-rpm induction motors. (b) Range of nominal efficiency for current industry energy-efficient TEFC 1800-rpm induction motors.

FIGURE 2.6 Comparisons of energy-efficient and standard motors. (Courtesy of MAGNETEK, St. Louis, MO.)

where

W_s = stator winding loss
W_r = rotor winding loss, slip loss
W_c = magnetic core loss
W_f = no-load friction and windage loss
W_{sl} = full-load stray load loss

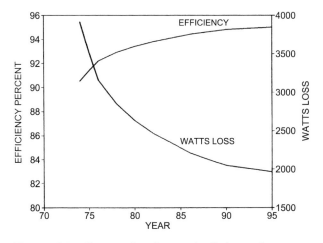

FIGURE 2.7 Loss reduction and efficiency improvement trend for 50-hp, 1800-rpm induction motor.

The accuracy of the efficiency determination depends on the test method used and the accuracy of the losses determined by the test method. There is no single standard method used throughout the industry. The most commonly referred to test methods are the following: IEEE Standard 112–1984 Standard Test Procedure for Polyphase Induction Motors and Generators; International Electrotechnical Commission (IEC) Publication 34-2, Methods of Determining Losses and Efficiency of Rotating Electrical Machinery from Tests; and Japanese Electrotechnical Commission (JEC) Standard 37 (1961), Standard for Induction Machines.

Each of these standards allows for more than one method of determining motor efficiency, and these can be grouped into two broad categories: direct measurement methods and segregated loss methods. In the direct measurement methods, both the input power and output power to the motor are measured directly. In the segregated loss methods, one or both are not measured directly. With direct measurement methods.

$$\text{Efficiency} = \frac{\text{output power}}{\text{input power} \times 100}$$

With segregated loss methods,

$$\text{Efficiency} = \frac{\text{input power} - \text{losses}}{\text{input power}} \times 100$$

or

$$\text{Efficiency} = \frac{\text{output power}}{\text{output power} + \text{losses}} \times 100$$

2.5.1 IEEE Standard 112–1984

Methods A, B, and C are direct measurement methods:

Method A: Brake. In this method, a mechanical brake is used to load the motor, and the output power is dissipated in the mechanical brake. The brake's ability to dissipate this power limits this method primarily to smaller sizes of induction motors (generally fractional horsepower).

Method B: Dynamometer. In this method, the energy from the motor is transferred to a rotating machine (dynamometer), which acts as a generator to dissipate the power into a load bank. The dynamometer is mounted on a load scale, a strain gauge, or a torque table. This is a very flexible and accurate test method for motors in the range 1–500 hp. However, to ensure accuracy, dynamometer corrections should be made as outlined in the test procedure. Method B includes a procedure for the stray load loss data smoothing by linear regression analysis. These smoothed values of stray load loss are used to calculate the final value of efficiency.

Method C: Duplicate Machines. This method uses two identical motors mechanically coupled together and electrically connected to two sources of power, the frequency of one being adjustable.* Readings are taken on both machines, and computations are made to calculate efficiency. This procedure includes a method of determining the stray load losses.

Methods E and F are segregated loss methods:

Method E: Input Measurements.[†] The motor output power is determined by subtracting the losses from the measure motor input power at different load points. For each load, the measured I^2R losses are adjusted for temperature and added to the no-load losses of friction, windage, and core. The stray load loss, which may be determined either directly, indirectly, or by the use of an agreed-on standardized value, is included in this total.

Method F: Equivalent Circuit Calculations. When load tests cannot be made, operating characteristics can be calculated from no-load and impedance data by means of an equivalent circuit. This equivalent circuit is shown in Fig. 2.8. Because of the nonlinear

---

* One machine is operated as a motor at rated voltage and frequency, and the other is driven as a generator at rated voltage per hertz but at a lower frequency to produce the desired load.

[†] In this method, it is necessary to connect the motor to a variable load. The input power is measured at the desired load points.

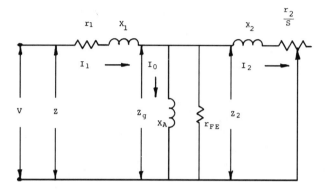

FIGURE 2.8 Polyphase induction motor per phase equivalent circuit. (From R. E. Osterlei, *Proceedings of the 7th National Conference on Power Transmission*, Gould Inc., St. Louis, MO, 1980.)

nature of these circuit parameters, they must be determined with great care to ensure accurate results. Procedures for determining these parameters are outlined in the standard as determined by a separate test. Accurate predictions of the motor characteristics depend on how closely r_2 represents the actual rotor resistance at low frequency.

2.5.2 IEC Publication 34-2

The same basic alternate methods as those outlined for IEEE 112 are also allowed for in IEC 34-2. However, a preference is expressed for the summation of losses method for the determination of motor efficiency. This is similar to IEEE 112 methods E and F except that the IEC method specifies stray load loss and temperature corrections differently. The IEC stray load losses are assumed to be 0.5% of rated input, whereas the IEEE standard states a preference for direct measurement of the stray load losses. The resistance temperature corrections in the IEC method are given as fixed values depending on insulation class, whereas the IEEE standard recommends use of the measured temperature rise for correcting resistance. These differences generally result in higher motor efficiency values by the IEC method.

2.5.3 JEC Standard 37

The JEC 37 standard also specifies the same basic methods as IEEE 112 with the exception of method *C*, duplicate machines. The preferred method for determining efficiency in this standard utilizes circle diagrams. This is a graphical solution of the *T* equivalent circuit of the induction motor. (This is similar to IEEE method F with the R_{fe} circuit branch.) As in the IEC standard, different methods are used to determine the circuit parameters and adjust the performance calculations. Principal among these is setting stray load loss equal to zero and using fixed values for resistance temperature corrections, which are a function of insulation class. These differences generally produce higher values for motor efficiency than the IEEE methods.

2.5.4 Comparison of Efficiencies Determined by Preferred Methods

To illustrate the variations in efficiency resulting from the use of the preferred methods, the full-load efficiency of several different polyphase motors was calculated by the preferred test methods given in the three standards. The results are shown in Table 2.5. As the values show, the efficiencies determined by the IEC and JEC methods are higher than the IEEE method. The major reason for this difference is the way in which stray load losses are accounted for. The IEEE method B stray load losses are included in the direct input and output measurements, whereas in the IEC method the stray load

TABLE 2.5 Efficiency Determined by Preferred Methods

hp	JEC 37, circle diagram	IEC 34-2 loss summation	IEEE 112, method B
5	88.8	88.3	86.2
10	89.7	89.2	86.9
20	91.9	91.4	90.4
75	93.1	92.7	90.0

Source: R. E. Osterlei, *Proceedings of the 7th National Conference on Power Transmission*, Gould Inc., St. Louis, MO, 1980.

losses are taken as 0.5% of the input, and in the JEC method they are set equal to zero. This comparison shows how important it is to know the method used to determine efficiency when comparing electric motor performance from different sources and countries.

2.5.5 Testing Variance

In addition to variance in efficiency due to test methods used, variances can also be caused by human error and test equipment accuracy. With dynamometer (IEEE 112, method B) testing, as with all test methods, there are several potential sources of inaccuracies: instrument accuracy, dynamometer accuracy, and instrument and dynamometer calibration. Therefore, to minimize these test errors, it is recommended that all the equipment and instruments be calibrated on a regular basis.

With proper calibration, dynamometer testing provides consistent and verifiable electric motor performance comparison. NEMA conducted a round-robin test of three different horsepower ratings (5, 25, and 100 hp) with a number of electric motor manufacturers. After a preliminary round of testing, each manufacturer was requested to test the motors in accordance with IEEE 112, method B, both with and without mathematical smoothing of the stray load loss. The results of these tests are summarized in Table 2.6.

TABLE 2.6 Variation in Test Data

	Variation in full-load efficiency			
	Without stray smoothing		With stray smoothing	
hp rating	Mean efficiency	Variation ± two std. deviations	Mean efficiency	Variation ± two std. deviations
5	86.3	2.0	87.1	0.7
25	89.6	1.3	89.5	0.8
100	92.7	1.3	91.9	0.9

Source: R. E. Osterlei, *Proceedings of the 7th National Conference on Power Transmission*, Gould Inc., St. Louis, MO, 1980.

Based on the test results, NEMA adopted a standard test procedure for polyphase motors rated 1–125 hp in accordance with IEEE Standard 112, method B, including mathematical smoothing of the stray load loss. It is recommended that this method of determining motor efficiency be used wherever possible.

2.6 MOTOR EFFICIENCY LABELING

Coincident with the NEMA test program, it was determined that a more consistent and meaningful method of expressing electric motor efficiency was necessary. The method should recognize that motors, like any other product, are subject to variations in material, manufacturing processes, and testing that cause variations in efficiency on a motor-to-motor basis for a given design.

No two identical units will perform in exactly the same way. Variance in the electrical steel used for laminations in the stator and rotor cores will cause variance in the magnetic core loss. Variance in the diameter and conductivity of the magnet wire used in the stator winding will change the stator winding resistance and hence the stator winding loss. Variances in the conductivity of aluminum and the quality of the rotor die casting will cause changes in the rotor power loss. Variances also occur in the manufacturing process. The quality of the heat treatment of the laminations for the stator and rotor cores can vary, causing a variance in the magnetic core loss. The winding equipment used to install the magnet wire in the stator can have tension that is too high, stretching the magnet wire and thus increasing the stator winding resistance and resulting in an increase in the stator winding loss $I^2 R$. Similarly, other variances, such as dimensional variances of motor parts, will contribute to the variation in motor efficiency.

It is a statistical fact that a characteristic of a population of a product will generally be distributed according to a bell-shaped or gaussian distribution curve. The height of the curve at any point is proportional to the frequency of occurrences, as illustrated in Fig. 2.9.

In the case of electric motors, the variation of losses for a population of motors of a given design is such that 97.7% of the motors will have an efficiency above the minimum efficiency defined

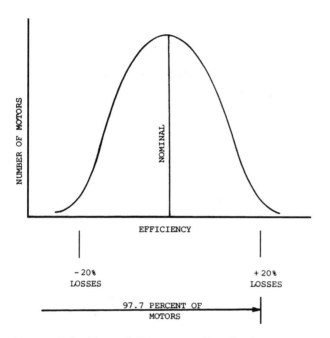

FIGURE 2.9   Normal frequency distribution.

by a variation of motor losses of ±20% of the losses at the nominal or average efficiency. Figure 2.10 illustrates the efficiency distribution for a specific value of nominal efficiency of 91%.

It is possible as motor manufacturers gain experience with this procedure that variations in losses will be lower than ±20%. In this event, the spread between nominal efficiency and minimum efficiency can be reduced.

Consequently, NEMA adopted a standard publication, MG1-12.54.2, recommending that polyphase induction motors be labeled with a NEMA nominal efficiency (or NEMA NOM EFF) when tested in accordance with IEEE Standard 112, dynamometer method, with stray loss smoothing. In addition, a minimum efficiency value was developed for each nominal efficiency value. Table 2.7 is a copy of the NEMA efficiency Table 12-6a. It is recommended that this method of labeling efficiency and testing be specified whenever possible.

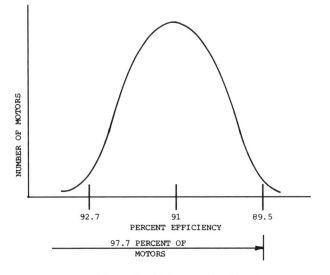

FIGURE 2.10 Normal efficiency distribution.

In instances in which guaranteed efficiencies are required, it is recommended that the preceding test method or an appropriate test method including the method of loss determination and the losses to be included in the efficiency determination be specified.

2.7 NEMA ENERGY-EFFICIENT MOTOR STANDARDS

In 1990, based on the experience gained by the electric motor manufacturers in producing energy-efficient polyphase induction motors and interest by industry, NEMA adopted a suggested standard for future design defining energy-efficient motors and setting efficiency levels for energy-efficient motors. These standards are as follows:*

MG1-2.43 Energy Efficient Polyphase Squirrel-cage Induction Motor. An energy efficient polyphase squirrel-

---

* Reprint by permission from NEMA Standards Publication No. MG1-1987, *Motors and Generators*, copyright 1987 by National Electrical Manufacturers Association.

TABLE 2.7 NEMA Nominal Efficiencies NEMA Table 12-6A

Nominal efficiency	Minimum efficiency	Nominal efficiency	Minimum efficiency
98.0	97.6	87.5	85.5
97.8	97.4	86.5	84.0
87.6	97.1	85.5	82.5
97.4	96.8	84.0	81.5
97.1	96.5	82.5	80.0
96.8	96.2	91.5	78.5
96.5	95.8	80.0	77.0
96.2	95.4	78.5	75.5
95.8	95.0	77.0	74.0
95.4	94.5	75.5	72.0
95.0	94.1	74.0	70.0
94.5	83.6	72.0	68.0
94.1	93.0	70.0	66.0
93.6	92.4	68.0	64.0
93.0	91.7	66.0	62.0
92.4	91.0	64.0	59.5
91.7	90.2	62.0	57.5
91.0	89.5	59.5	55.0
90.2	88.5	57.5	52.5
89.5	87.5	55.0	50.5
88.5	86.5	52.5	48.0

Source: Reprinted by permission from NEMA Standard Publication No. MG1-1987, *Motors and Generators*, copyright 1987 by the National Electrical Manufacturers Association.

cage induction motor is one having an efficiency in accordance with MG1-12.55A. Suggested Standard for Future Design.

MG1-12.55A Efficiency Levels of Energy Efficient Polyphase Squirrel-cage Induction Motors. The nominal full-load efficiency determined in accordance with MG1-12.54.1 and identified on the nameplate in accordance with MG1-12.54.2 shall equal or exceed the values listed

TABLE 2.8 NEMA Table 12-6c Full-Load Nominal Efficiencies and Associated Minimum Efficiencies for Polyphase Induction Motors

hp	2-Pole Nom.	2-Pole Min.	4-Pole Nom.	4-Pole Min.	6-Pole Nom.	6-Pole Min.	8-Pole Nom.	8-Pole Min.
1.0	—	—	82.5	81.5	80.0	78.5	74.0	72.0
1.5	82.5	81.5	84.0	82.5	84.0	82.5	75.5	74.0
2.0	84.0	82.5	84.0	82.5	85.5	84.0	85.5	84.0
3.0	84.0	82.5	86.5	85.5	86.5	85.5	86.5	85.5
5.0	85.5	84.0	87.5	86.5	87.5	86.5	87.5	86.5
7.5	87.5	86.5	88.5	87.5	88.5	87.5	88.5	87.5
10.0	88.5	87.5	89.5	88.5	90.2	89.5	89.5	88.5
15.0	89.5	88.5	91.0	90.2	90.2	89.5	89.5	88.5
20.0	90.2	89.5	91.0	90.2	91.0	90.2	90.2	89.5
25.0	91.0	90.2	91.7	91.0	91.7	91.0	90.2	89.5
30.0	91.0	90.2	92.4	91.7	92.4	91.7	91.0	90.2
40.0	91.7	91.0	93.0	92.4	93.0	92.4	91.0	90.2
50.0	92.4	91.7	93.0	92.4	93.0	92.4	91.7	91.0
60.0	93.0	92.4	93.6	93.0	93.6	93.0	92.4	91.7
75.0	93.0	92.4	94.1	93.6	93.6	93.0	93.6	93.0
100.0	93.0	92.4	94.1	93.6	94.1	93.6	93.6	93.0
125.0	93.6	93.0	94.5	94.1	94.1	93.6	93.6	93.0
150.0	93.6	93.0	95.0	94.5	94.5	94.1	93.6	93.0
200.0	94.5	94.1	95.0	94.5	94.5	94.1	93.6	93.0
Enclosed Motors								
1.0	75.5	74.0	82.5	81.5	80.0	78.5	74.0	72.0
1.5	82.5	81.5	84.0	82.5	85.5	84.0	77.0	75.5
2.0	84.0	82.5	84.0	82.5	86.5	85.5	82.5	81.5
3.0	85.5	84.0	87.5	86.5	87.5	86.5	84.0	82.5
5.0	87.5	86.5	87.5	86.5	87.5	86.5	85.5	84.0
7.5	88.5	87.5	89.5	88.5	89.5	88.5	85.5	84.0
10.0	89.5	88.5	89.5	88.5	89.5	88.5	88.5	87.5
15.0	90.2	89.5	91.0	90.2	90.2	89.5	88.5	87.5
20.0	90.2	89.5	91.0	90.2	90.2	89.5	89.5	88.5
25.0	91.0	90.2	92.4	91.7	91.7	91.0	89.5	88.5
30.0	91.0	90.2	92.4	91.7	91.7	91.0	91.0	90.2
40.0	91.7	91.0	93.0	92.4	93.0	92.4	91.0	90.2
50.0	92.4	91.7	93.0	92.4	93.0	92.4	91.7	91.0

TABLE 2.8 Continued

hp	2-Pole Nom.	2-Pole Min.	4-Pole Nom.	4-Pole Min.	6-Pole Nom.	6-Pole Min.	8-Pole Nom.	8-Pole Min.
60.0	93.0	92.4	93.6	93.0	93.6	93.0	91.7	91.0
75.0	93.0	92.4	94.1	93.6	93.6	93.0	93.0	92.4
100.0	93.6	93.0	94.5	94.1	94.1	93.6	93.0	92.4
125.0	94.5	94.1	94.5	94.1	94.1	93.6	93.6	93.0
150.0	94.5	94.1	95.0	94.5	95.0	94.5	93.6	93.0
200.0	95.0	94.5	95.0	94.5	95.0	94.5	94.1	93.6

Source: Reprinted by permission from NEMA Standards Publication No. MG1-1987, *Motors and Generators*, copyright 1987 by the National Electrical Manufacturers Association.

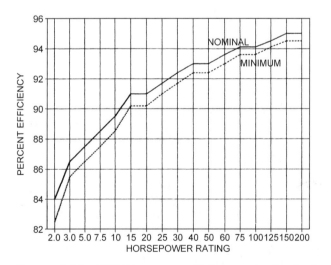

FIGURE 2.11 NEMA energy-efficiency standards for four-pole open induction motors from data in Table 2.8. (Courtesy National Electrical Manufacturers Association, Washington, DC.)

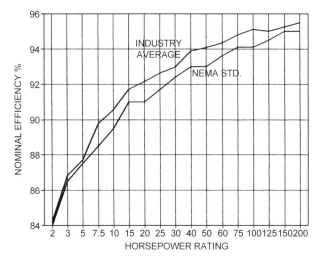

FIGURE 2.12 Comparison of NEMA nominal efficiency and available industry average efficiency for 1800-rpm open energy-efficient motors.

in Table 12-6c for the motor to be classified as "energy-efficient." Suggested Standard for Future Design.

As mentioned earlier, the variation in the nominal efficiency of energy-efficient induction motors has been smaller than for standard induction motors. In addition, the electric motor manufacturers have experienced less efficiency variation in their product. This is reflected in the NEMA standard for energy-efficient motors, in that the variation in the allowable losses has been reduced to ±10%, which means a higher minimum efficiency for a given nominal efficiency. Table 2.8 is a copy of the NEMA Table 12-6C with the higher full-load nominal and minimum efficiency standards for energy-efficient motors, both open and TEFC, at various speeds. Figure 2.11 shows the relationship between the NEMA nominal efficiency and the minimum efficiency for energy-efficient four-pole open motors. Figure 2.12 is a comparison of the NEMA nominal efficiencies and the average of the available industry energy-efficient four-pole open motors. This indicates that available energy-efficient motors have efficiencies slightly higher than the NEMA standard.

3

Fundamentals of Electric Motor Drives

Electric machines are an essential part of industry. They provide the necessary mechanical-to-electrical or electrical-to-mechanical conversion. In the United States, more than 50% of the electric power is consumed by electric motors. The motors perform many different functions, from small applications like cooling fans in your personal computer that consume only a few watts of power all the way to huge pumps that consume megawatts.

The majority of motors used today, approximately 80%, are three-phase induction motors. Motors themselves have limited capabilities. In more complicated and technical applications, the motor itself does not perform the necessary tasks. Today, a complete electric drive system is needed to control and manipulate the motor to fit specific applications. An electric drive system involves the control of electric motors in steady-state and dynamic operations. The system should take into account the type of mechanical load.

There are many different types of mechanical loads. When defining a load, torque versus speed characteristics are explored.

The relationship is then used to design the type of electric drive system. In the past, electric drive systems required large and expensive equipment. These systems were inefficient and limited to the specific applications they where designed for.

Today, with advancements in power electronics, control electronics, microprocessors, microcontrollers, and digital signal processors (DSPs), electric drive systems have improved drastically. Power electronic drives are more reliable, more efficient, and less expensive. In fact, a power electronic drive on average consumes 25% less energy than a classic motor drive system. The advancements in solid-state technologies are making it possible to build the necessary power electronic converters for electric drive systems.

The power electronic devices allow motors to be used in more precise applications. Such systems may include highly precise speed or position control. Systems that used to be controlled pneumatically and hydrolically can now be controlled electrically as well.

3.1 POWER ELECTRONIC DEVICES

The *power diode* is the simplest, uncontrollable power electronic switch. A power diode is forward biased (on) when its current is positive and reverse biased (off) when its voltage is negative.

A *thyristor* is a controllable three-terminal device. If a current pulse is applied to its gate, the thyristor can be turned on and conducts current from its anode to cathode, providing a positive anode-to-cathode voltage. However, in order to turn a thyristor on, the gate current must be above a minimum value I_{GT}. After the thyristor turns on, if its current (i.e., anode to cathode) reaches above a minimum value called latching current I_L, the gate current is no longer required. The thyristor will continue to conduct until its current falls below a minimum value called holding current I_H.

A *diac* is a two-terminal power electronic device. When the voltage across the terminals reaches the diac specific voltage, the diac is turned on and conducts current from the positive terminal to the negative terminal. The voltage across terminals decreases to a small value which is the voltage drop while diac is on. A diac is a bidirectional device.

A *triac* is a three-terminal, controllable power electronic switch. The operation of a triac is equivalent to two thyristors which are parallel in opposite directions. Therefore, a triac has the capability of conducting current in both directions. Gate current can also be positive or negative. As a result, a triac has four different operating modes.

Power transistors have the characteristics of conventional transistors. However, they have the capability of conducting higher collector current. They have also higher breakdown voltage V_{CEO}. Power transistors are designed for high-current, high-voltage, and high-power applications. They are usually operated either in the fully on or fully off state.

Power MOSFETs are voltage-controlled devices. They are usually N-channel and of the enhancement type. Most power MOSFETs are off when $V_{GS} < 2$ V and are on when $V_{GS} > 4$ V. When a power MOSFET is on there is a small resistance, i.e., less than 1 Ω, between drain and source, and when it is off there is a large resistance (almost open circuit) between drain and source.

Isolated gate bipolar transistors (IGBTs) are equivalent to power transistors whose bases are driven by MOSFETs. Similar to a MOSFET, an IGBT has a high impedance gate, which requires only a small amount of energy to switch the device. Like a power transistor, an IGBT has a small on-state voltage.

Unijunction transistors (UJTs) are three-terminal devices with one emitter and two bases. It can be assumed that between the two bases two resistors are connected in series. If in the forward-biased mode the emitter voltage reaches the voltage divided between the two resistors, emitter current suddenly increases and the device conducts.

Pulse transformers are similar to conventional transformers. However, they are designed for high-frequency and very low-power applications. They are not for transforming power from the primary to the secondary. They are used for isolating control circuits from power circuits in power electronic applications.

3.2 ELECTRIC MOTOR DRIVES

An electric motor drive system is made up of five main components. Figure 3.1 shows a block diagram of an electric motor drive. The input to the drive is the power source. The power source is the energy for the system. Next is the power electronic converter. The electronic converter manipulates the voltage, current, and frequency provided by the power source. To control the system, a controller is needed. Of course, the remaining two components of the system are a motor and a mechanical load.

In the following sections, different types of electric motor drive systems are explored. Drives for DC and AC motors are explained in detail.

3.3 SINGLE-PHASE, HALF-WAVE, CONTROLLED RECTIFIER

Shown in Fig. 3.2 is a separately excited DC motor controlled by a single-phase, half-wave controlled rectifier. This rectifier provides speed control for the separately excited DC motor by varying armature voltage and current.

The steady state voltage and torque equations for a separately excited DC motor are

$$V_a = R_a I_a + E_a$$

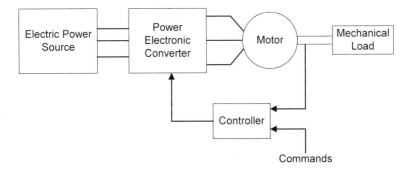

FIGURE 3.1 Block diagram of an electric motor drive.

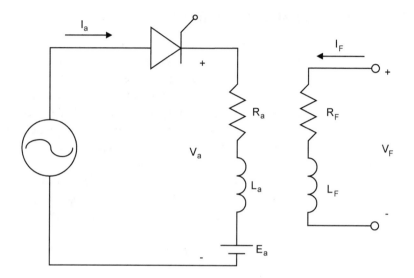

FIGURE 3.2 Single-phase, half-wave, controlled rectifier.

$$V_F = R_F I_F$$
$$E_a = K I_F \omega$$
$$T = K I_F I_a$$

Combining these equations, we get

$$\omega = \frac{V_a - R_a I_a}{K I_F}$$

or

$$\omega = \frac{V_a}{K I_F} - \frac{R_a T}{(K I_F)^2}$$

As seen from these equations, speed control can be done by varying R_a, V_a, and I_F. Varying R_a can only increase the speed. Also, by varying R_a, we increase the losses $I^2 R$. IF can only be decreased. This is accomplished by putting resistance in the field. Again, this increases the $I^2 R$ losses. Control of V_F is limited because of saturation. The best method of controlling the speed of the motor is to control V_a. Using a thyristor, as shown in Fig. 3.2, the average of V_a can be controlled. Figure 3.3 shows the voltage and current for a separately excited DC motor controlled by a single-phase, half-wave, controlled rectifier. As seen in Fig. 3.3, α represents the firing angle of the thyristor. This is where the thyristor turns on. Also seen in Fig. 3.3 is β. β is where the current I_a reaches zero and the thyristor turns off. From these two quantities, we define the conduction angle:

$$\gamma = \beta - \alpha$$

By changing the conduction angle, the average value of V_a is varied. When V_a is varied, the speed of the motor is changed. This is a simple but effective speed controller.

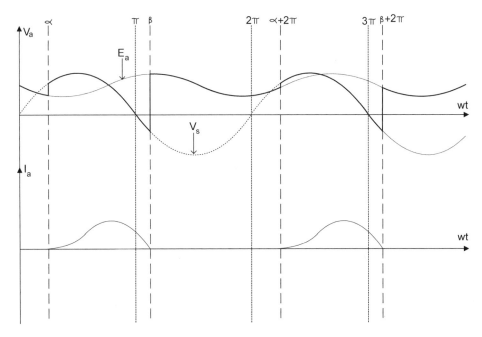

FIGURE 3.3 Voltage and current for Fig. 3.2.

3.4 SINGLE-PHASE, FULL-WAVE, CONTROLLED RECTIFIER

The next electric drive discussed is a single-phase, full-wave controlled rectifier. This adjustable speed drive is similar to the single-phase, half-wave controlled rectifier. As an example, this rectifier is presented to control a separately excited DC motor. The single-phase, full-wave controlled rectifier consists of four thyristors. The increase in thyristors provides for better control compared to the half-wave controlled rectifier. The obvious disadvantage of the full-wave rectifier is the increase in price because of the increase in the number of thyristor. Figure 3.4 shows a separately excited DC motor controlled by a single-phase, full-wave controlled rectifier.

There are three different modes of operation for the single-phase, full-wave controlled rectifier. The first is discontinuous conduction mode (DCM). In DCM, the current I_a reaches zero and stays at zero for a certain period of time. The next mode is continuous conduction mode (CCM). In CCM, I_a does not reach

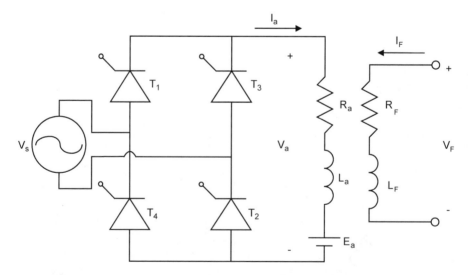

FIGURE 3.4 Single-phase, full-wave, controlled rectifier.

zero at any point during the period. The finial mode of operation is critically discontinuous conduction mode (CDCM). In CDCM, the current I_a reaches zero and then immediately starts to increase.

Unlike the half-wave rectifier, the full-wave rectifier has the ability to manipulate the current when V_s is negative. There are three modes in DCM. The first mode is when from time $t = 0$ until α. Mode one is shown in Fig. 3.5. In mode one, there is no current in the armature; this results in a V_a equal to the back emf E_a.

Mode two occurs when the source voltage V_s is positive. Mode two is shown in Fig. 3.6. In mode two, T_1 and T_2 are conducting and T_3 and T_4 are not conducting. V_a is equal to the source voltage V_s.

The finial mode is mode three. Mode three is shown in Fig. 3.7. This mode is the opposite of mode two. In mode three, T_3 and T_4 are conducting and T_1 and T_2 are not conducting. This makes the voltage V_a equal to the negative of the source voltage V_s.

Figure 3.8 shows the waveforms of the rectifier in DCM. As seen in Fig. 3.8, the conducting angle occurs twice per period, once when the source voltage V_s is positive and again when the source

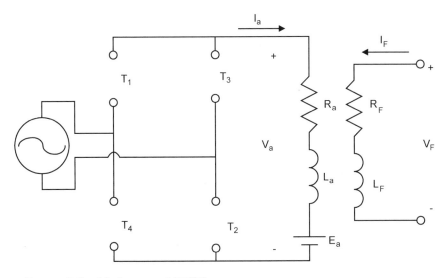

FIGURE 3.5 Mode one of DCM.

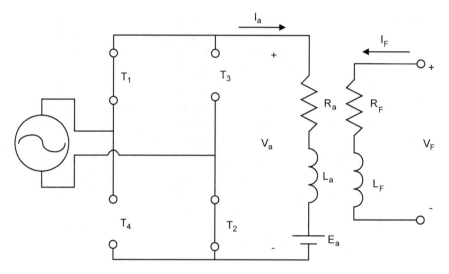

FIGURE 3.6 Mode two of DCM.

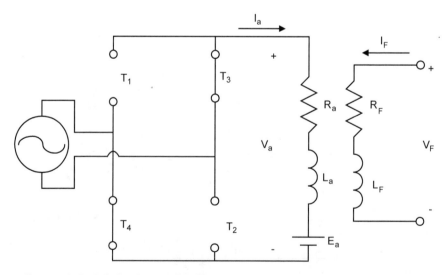

FIGURE 3.7 Mode three of DCM.

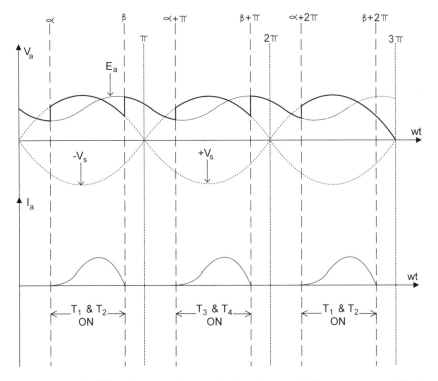

FIGURE 3.8 Waveforms of a single-phase, full-wave controlled rectifier in DCM.

voltage is negative. In addition, in mode one, the voltage V_a is equal to the back EMF E_a until α.

Continuous conduction mode is similar to DCM, but in CCM mode one does not exist. Figure 3.9 shows the circuit operating in CCM. CCM occurs when V_a is large compared to E_a. Notice that V_a is never equal to E_a.

3.5 PHASE-CONTROLLED INDUCTION MOTOR DRIVES

As mentioned earlier, three-phase induction motors make up the majority of the motors in the industry. This is because of their low

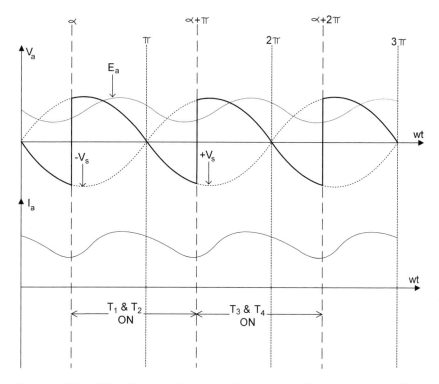

FIGURE 3.9 Waveforms of a single-phase, full-wave controlled rectifier in CCM.

cost, low maintenance, and high efficiency. To control any motor, a strong understanding of the motor equations is needed. Before discussing the phase-controlled induction motor drives, the necessary equations are reviewed.

An induction motor is very similar to a transformer. The induction motor is an AC machine in which alternating current is supplied to the stator directly and to the rotor windings by induction, like a transformer. There are two types of induction motors: a wound-rotor induction motor and a squirrel-cage rotor induction motor. The wound rotor has slip rings mounted on the rotor shaft. These slip rings enable the user to access the rotor winding. For steady-state operation, these windings are shorted. A squirrel-cage rotor has conducting bars embedded in slots in the rotor magnetic core. These bars are short-circuited at each end.

The synchronous speed depends on the number of pole pairs and the frequency of the source. Synchronous speed (N_s) in revolutions per second is given by

$$N_s = \frac{f}{P}$$

where f is the frequency of the source and P is the number of pole pairs. The difference in rotor speed N and synchronous speed N_s is represented by slip s, which is

$$s = \frac{N_s - N_r}{N_s}$$

At full load, slip is in the range of 2–3%. A very important characteristic of the induction motor is the speed-torque. To develop the torque equation of the induction motor the approximate equivalent circuit of an induction motor is used. Figure 3.10 shows the approximate equivalent circuit of an induction machine.

The power transferred from the stator to the rotor is called the air gap power P_g. The air gap power is the product of the torque T and the synchronous speed ω_s in radians. Using the equivalent circuit, the air gap power can be represented with motor parameters. The air gap power is defined as

$$P_g = T\omega_s$$

$$P_g = \frac{R_2'}{s} I_2'$$

Mechanical power P_m is the product of torque and rotor speed ω. Also using the equivalent circuit, the mechanical power can be represented using the motor parameters. P_m is represented by

$$P_m = T\omega$$

$$P_m = R_2'\left(\frac{1}{s} - 1\right) I_2'$$

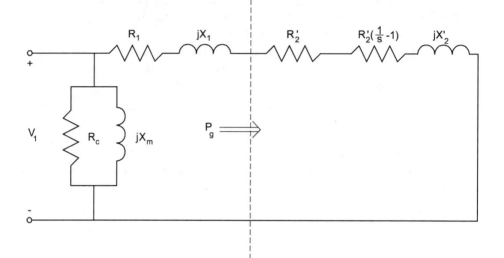

FIGURE 3.10 Approximate equivalent circuit of an induction machine.

Combining these equations, torque can be represented by

$$T = \frac{P_g}{\omega_s} = \frac{(R_2'/s)I_2^2}{\omega_s}$$

and

$$T = \frac{P_m}{\omega} = \frac{R_2'(1/s - 1)I_2'}{\omega}$$

Using the equivalent circuit and the above equations, torque is represented as a function of voltage and speed:

$$T = \frac{R_2'/s}{\omega_s}I_2' = \frac{V_1^2}{\omega_s}\frac{R_2'/s}{(R_1 + R_2'/s)^2\ (X_1 + X_2')^2}$$

This equation represents the torque of an induction motor as a function of voltage and speed. If the voltage is varied, the torque of the motor changes. If the frequency is varied, the speed of the motor changes. Saturation will occur if the voltage is very large or the frequency is very low. To obtain optimal flux operation, V/f should be kept constant. This will not make the motor go into saturation.

A very simple phase-controlled induction motor is explored. To control a three-phase induction motor, an AC/AC converter can be used. Figure 3.11 shows a three-phase AC/AC converter. Notice that each phase has two thyristors. The firing angles of the thyristors in each phase are 180 degrees apart and are positioned opposite of each other. In addition, as expected, each phase is 120 degrees apart. To simplify the analysis of the phase-controlled induction motor drives, only one phase of the AC/AC converter is examined. Figure 3.12 shows phase A of the converter.

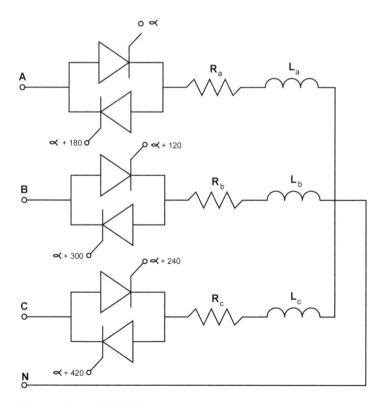

FIGURE 3.11 AC/AC converter.

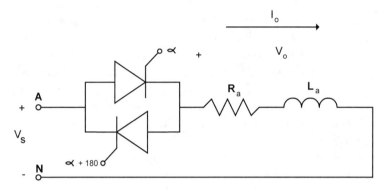

FIGURE 3.12 Phase A of the AC/AC converter.

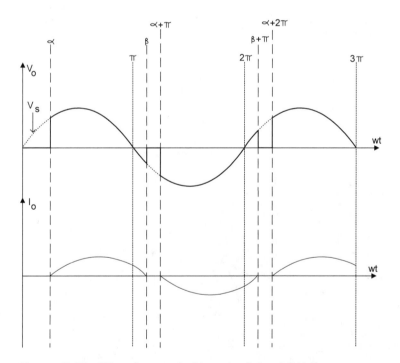

FIGURE 3.13 Waveforms of phase A of the AC/AC converter.

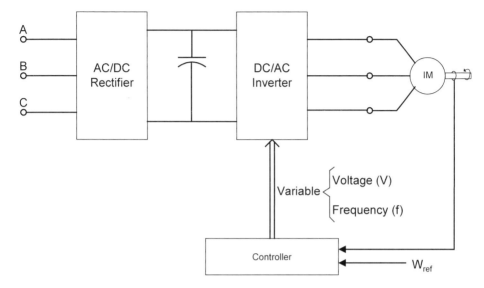

FIGURE 3.14   Block diagram of an advanced induction motor drive.

The waveforms of the circuit of Fig. 3.12 are shown in Fig. 3.13. This is not the most effective way of controlling an induction motor. The quality of the output voltage is not good. Furthermore, as mentioned earlier, to obtain the optimal flux operation V/f should be kept constant. Using an AC/AC converter does not keep V/f constant. A more effective induction motor drive is shown in Fig. 3.14. A constant voltage and frequency three-phase source is the input. With the AC/DC rectifier, DC/AC inverter, and controller, the voltage and frequency can be varied. The control can be used to keep V/f constant and provide optimal flux operation.

3.6   CONTROL OF DC MOTORS USING DC/DC CONVERTERS

Using DC/DC converters to control DC motors is very effective. The speed of the motor is controlled by the on and off time of the DC voltage. This is done through different switching schemes. The switching schemes are varied to produce the control needed.

To understand the switching schemes, one must first understand some definitions. Figure 3.15 shows a typical voltage graph produced by a DC/DC converter. It shows a series of pulses. On the graph, the on time t_{on} and off time t_{off} are defined. From these quantities, the period T and duty cycle d are defined as follows.

$$T = t_{on} + t_{off}$$

$$d = \frac{t_{on}}{t_{on} + t_{off}} = \frac{t_{on}}{T}$$

Using these definitions, different switching schemes can be defined. There are two main switching schemes. The first is pulse width modulation (PWM). In PWM, the period is constant and the duty cycle is variable. However, the duty cycle is limited between zero and one; therefore, the on time is less than the period. The second switching scheme is frequency modulation (FM). Within FM, there are two basic conditions. One is where t_{on} is constant and the period is varied. Another is when t_{off} is constant and the period is varied. Figure 3.16 shows the different switching schemes. Comparing the two different switching schemes, PWM has fewer harmonics and produces less noise.

Figure 3.17 shows a DC separately excited motor controlled by a DC/DC converter. To simplify the analysis, the switch Q will be assumed to be ideal. Another assumption is that E_a will be considered constant. This is a reasonable assumption because I_F will be

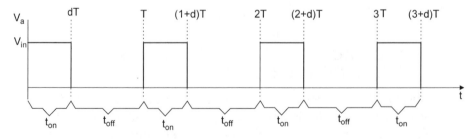

FIGURE 3.15 Typical voltage graph.

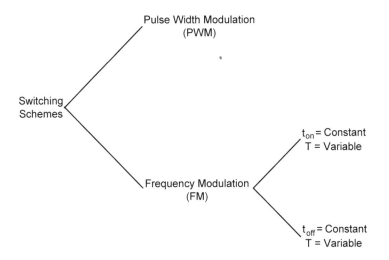

FIGURE 3.16 Switching schemes for DC/DC converter.

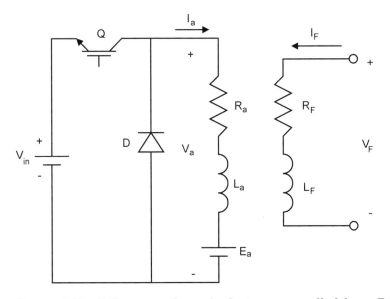

FIGURE 3.17 DC separately excited motor controlled by a DC/DC converter.

constant and we can assume ω is constant. Figure 3.18 shows the
output currents produced by the DC/DC converter controlling a
DC separately excited motor in continuous conduction mode
(CCM).

To find the maximum current, shown in Fig. 3.18, the circuit in
Fig. 3.19 will be analyzed. During this time $0 < t < dT$, V_a is equal to
$V_{in} - E_a$, and $i_a(t=0) = I_{a,min}$. Using simple circuit analysis on the
circuit in Fig. 3.19, we find that $i_a(t)$ is

$$i_a(t) = \frac{V_{in} - E_a}{R_a}\left(1 - e^{\frac{-R_a t}{L_a}}\right) + I_{a,min}e^{\frac{-R_a}{L_a}t}$$

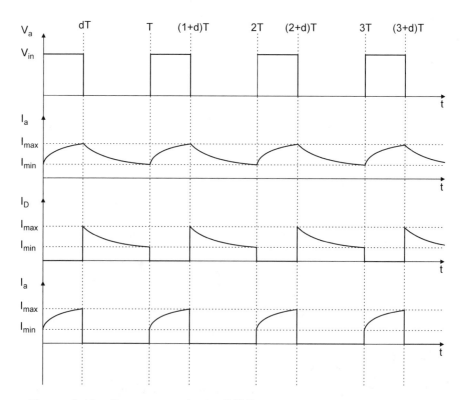

FIGURE 3.18 Current graphs in CCM.

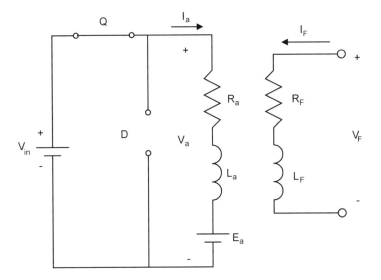

FIGURE 3.19 Continuous conduction mode when $0 < t < dT$.

Now, as seen from Fig. 3.18, $i_a(t = dT) = I_{a,max}$; therefore,

$$I_{a,max} = \frac{V_{in} - E_a}{R_a}\left(1 - e^{\frac{-R_a}{L_a}dT}\right) + I_{a,min}e^{\frac{-R_a}{L_a}dT}$$

Figure 3.20 shows the circuit during $dT < t < T$. During this time interval, Q is off and D is conducting.

Using the fact that V_a is equal to zero and $i_a(t = dT) = I_{a,max}$, we can find the current. From Fig. 3.20, the current is

$$i_a(t) = \frac{-E_a}{R_a}\left(1 - e^{\frac{-R_a}{L_a}(t-dT)}\right) + I_{a,max}e^{\frac{-R_a}{L_a}(t-dT)}$$

Now, as seen from Fig. 3.18, $i_a(t = T) = I_{a,min}$; therefore,

$$I_{a,min} = \frac{-E_a}{R_a}\left(1 - e^{\frac{-R_a}{L_a}(t-d)T}\right) + I_{a,max}e^{\frac{-R_a}{L_a}(t-d)T}$$

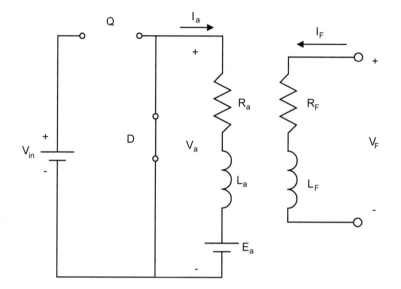

FIGURE 3.20   Continuous conduction mode when $dT < t < T$.

From these equations, the relations for $I_{a,min}$ and $I_{a,max}$ are found as follows:

$$I_{a,min} = \frac{V_{in}}{R_a}\frac{e^{\frac{R_a}{L_a}dT} - 1}{e^{\frac{R_a}{L_a}T} - 1} - \frac{E_a}{R_a}$$

$$I_{a,max} = \frac{V_{in}}{R_a}\frac{1 - e^{\frac{-R_a}{L_a}dT}}{1 - e^{\frac{-R_a}{L_a}T}} - \frac{E_a}{R_a}$$

In order to find the relationship between the duty cycle and voltage, the average V_a must be found:

$$\langle V_a \rangle = \frac{1}{T}\int_0^T V_a dt = \frac{1}{T}\int_0^{dT} V_{in} dt = d\ V_{in}$$

The speed of the DC separately excited motor can be changed by changing the duty cycle. If the inductance of the motor is too small, the current will go into discontinuous conduction mode. Figure 3.21 shows the voltage and current waveforms for DCM.

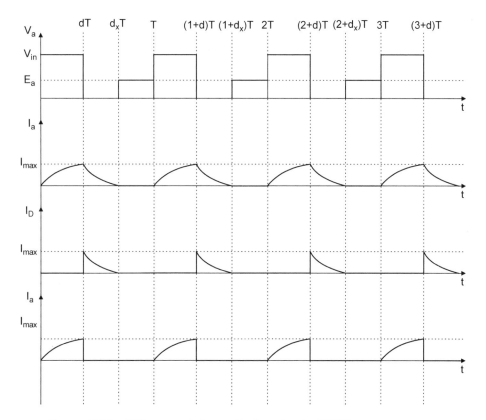

FIGURE 3.21 DC separately excited motor in DCM.

To perform the analysis for DCM, different modes of the circuit will be analyzed. Figure 3.22 shows the circuit for $0 < t < dT$. During $0 < t < dT$, V_a is equal to $V_{in} - E_a$ and $i_a(t=0) = 0$. Therefore,

$$i_a(t) = \frac{V_{in} - E_a}{R_a} \left(1 - e^{\frac{-R_a t}{L_a}}\right)$$

In DCM, $I_{a,min}$ is equal to zero and $I_{a,max}$ is simply equal to

$$I_{a,max} = \frac{V_{in} - E_a}{R_a} \left(1 - e^{\frac{-R_a}{L_a}dT}\right)$$

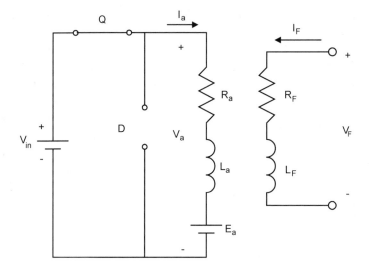

FIGURE 3.22 Discontinuous conduction mode during $0 < t < dT$.

From Fig. 3.21, we see that during the time $dT < t < d_x T$, the current is decreasing until it reaches zero. Figure 3.23 shows the circuit during this time interval.

The equation for $i_a(t)$ for the time shown in Fig. 3.23 is

$$i_a(t) = \frac{-E_a}{R_a}\left(1 - e^{\frac{-R_a}{L_a}(t-dT)}\right) + I_{a,\max} e^{\frac{-R_a}{L_a}(t-dT)}$$

To find the value of d_x, the above equation can be solved for the condition $i_a(t = d_x T) = 0$. The final portion of the graph shown in Fig. 3.21 to analyze is $d_x T < t < T$. Figure 3.24 shows the circuit during $d_x T < t < T$. During this time interval, Q and D are not conducting. This makes $i_a = 0$ and $V_a = E_a$.

In conclusion, electric motor drives have advanced since the old mechanically linked systems. The new drives are more accurate and consume less power. For example, slowing a pump or fan by using an electric drive reduces energy consumption more effectively than allowing the motor to run at constant speed and then restricting or bypassing the flow with a valve or damper.

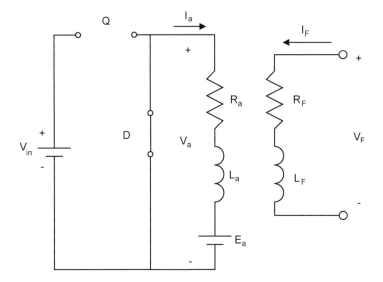

FIGURE 3.23 Discontinuous conduction mode when $dT < t < d_x T$.

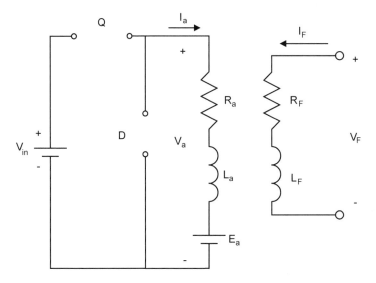

FIGURE 3.24 Discontinuous conduction mode when $d_x T < t < T$.

In addition, electric motor drives can provide benefits when starting a motor. They can be used to slowly start the motor. The slow start reduces the mechanical stress on the motor and the load equipment. By slowly starting a motor, the voltage sag is reduced. Voltage sag causes lights to dim and other equipment like computers to shut down.

The future of electric motor drives is to have a simple motor and a complex power electronic converter. The power electronic converter is software based, while the electric machine is hardware based. Software is easier to manipulate than hardware. Furthermore, from a manufacturing standpoint, it is less expensive to have a software intensive drive compared to a hardware intensive drive. With the majority of electric power consumed by electric motors, the future of electric motor drives is good.

SELECTED READINGS

1. Krishnan, R. (2001). *Electric Motor Drives: Modeling, Analysis, and Control*. Upper Saddle River, NJ: Prentice-Hall.
2. Mohan, N., Undeland, T. M., Robbins, W. P. (2003). *Power Electronics: Converters, Applications, and Design*. New York: John Wiley & Sons.
3. El-Sharkawi, A. (2000). *Fundamentals of Electric Drives*. Pacific Grove, PA: Brooks/Cole Publishing.
4. Bose, B. K. (2002). *Modern Power Electronics and AC Drives*. Prentice Hall PTR.
5. Mohan, N. (2001). *Electric Drives: An Integrative Approach*. Minneapolis: MNPERE.
6. Mohan, N. (2001). *Advanced Electric Drives*. Minneapolis: MNPERE.
7. Skvarenina, T. L. (2002). *The Power Electronics Handbook*. Boca Raton, FL: CRC Press.
8. Kassakian, J. G., Schlecht, M. F., Verghese, G. C. (1991). *Principles of Power Electronics*. Upper Saddle River, NJ: Addison Wesley.
9. Krein, P. T. (1998). *Elements of Power Electronics*. New York: Oxford University Press.

10. Rashid, M. H. (2003). *Power Electronics*. 3rd ed. Upper Saddle River, NJ: Prentice Hall.
11. Hart, D. H. (1997). *Introduction to Power Electronics*. Upper Saddle River, NJ: Prentice-Hall.
12. Erickson, R. W., Maksimovic, D. (2001). *Fundamentals of Power Electronics*. 2nd ed. Norwell, MA: Kluwer Academic Publishers.
13. Trzynadlowski, A. M. (1998). *Introduction to Modern Power Electronics*. New York: John Wiley & Sons.
14. Trzynadlowski, A. M. (1994). *The Field Orientation Principle in Control of Induction Motors*. Norwell, MA: Kluwer Academic Publishers.
15. Vas, P. (1990). *Vector Control of AC Machines*. New York: Oxford Science Publications.
16. Rajashekara, K., Kawamura, A., Matsuse, K. (1996). *Sensorless Control of AC Motor Drives: Speed and Position Sensorless Operation*. Piscataway, NJ: IEEE Press.

4

The Power Factor

The advantages of improving the equipment and system power factor are not as obvious as the advantages of improving the kilowatt power consumption. Improvement in the plant power factor can result in savings in kilowatt-hour power consumption due to lower distribution and transformer losses and, in many cases, a substantial reduction in the energy demand charge.

4.1 WHAT IS THE POWER FACTOR?

Traditionally, power factor has been defined as the ratio of the kilowatts of power divided by the kilovolt-amperes drawn by a load or system, or the cosine of the electrical angle between the kilowatts and kilovolt-amperes. However, this definition of power factor is valid only if the voltages and currents are sinusoidal. When the voltages and/or currents are nonsinusoidal, the power factor is reduced as a result of voltage and current harmonics in the system.

Therefore, the discussion of power factor will be considered for the two categories, i.e., systems in which the voltages and currents are substantially sinusoidal and systems in which the voltages and currents are nonsinusoidal as a result of nonlinear loads.

4.2 THE POWER FACTOR IN SINUSOIDAL SYSTEMS

The line current drawn by induction motors, transformers, and other inductive devices consists of two components: the magnetizing current and the power-producing current.

The magnetizing current is that current required to produce the magnetic flux in the machine. This component of current creates a reactive power requirement that is measured in kilovolt-amperes reactive (kilovars, kvar). The power-producing current is the current that reacts with the magnetic flux to produce the output torque of the machine and to satisfy the equation

$$T = K\Phi I$$

where

T = output torque
Φ = net flux in the air gap as a result of the magnetizing current
I = power-producing current
K = output coefficient for a particular machine

The power-producing current creates the load power requirement measured in kilowatts (kW). The magnetizing current and magnetic flux are relatively constant at constant voltage. However, the power-producing current is proportional to the load torque required.

The total line current drawn by an induction motor is the vector sum of the magnetizing current and the power-producing current. For three-phase motors, the apparent power, or kilovolt-ampere (kVA) input to the motor, is

$$\text{kVA} = \frac{I_L V_L \sqrt{3}}{1000}$$

where

I_L = total line current
V_L = line-to-line voltage

The vector relationship between the line current I_L and the reactive component I_x and load component I_p currents can be expressed by a vector diagram, as shown in Fig. 4.1, where the line current I_L is the vector sum of two components. The power factor is then the cosine of the electrical angle θ between the line current and phase voltage.

This vector relationship can also be expressed in terms of the components of the total kilovolt-ampere input, as shown in Fig. 4.2. Again, the power factor is the cosine of the angle θ between the total kilovolt-ampere and kilowatt inputs to the motor. The kilovolt-ampere input to the motor consists of two components: load power, i.e., kilowatts, and reactive power, i.e., kilovars.

The system power factor can be determined by a power factor meter reading or by the input power (kW), line voltage, and line current readings. Thus,

$$\text{Power factor} = \frac{\text{kW}}{\text{kVA}}$$

where

$$\text{kVA} = \frac{I_L V_L \sqrt{3}}{1000}$$

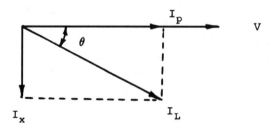

FIGURE 4.1 Vector diagram of load current for one phase of the motor.

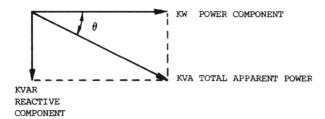

FIGURE 4.2 Vector diagram of power input without a power factor correction.

Then kvar is

$$kvar = \sqrt{(kVA)^2 - (kW)^2}$$

An inspection of the kilovolt-ampere input diagram shows that the larger the reactive kilovar, the lower the power factor and the larger the kilovolt-ampere for a given kilowatt input.

4.3 WHY RAISE THE POWER FACTOR?

A low power factor causes poor system efficiency. The total apparent power must be supplied by the electric utility. With a low power factor, or a high-kilovar component, additional generating losses occur throughout the system. Figures 4.3 and 4.4 illustrate the effect of the power factor on generator and transformer capacity. To discourage low–power factor loads, most utilities impose some form of penalty or charge in their electric power rate structure for a low power factor.

When the power factor is improved by installing power capacitors or synchronous motors, several savings are made:

1. A high power factor eliminates the utility penalty charge. This charge may be a separate charge for a low

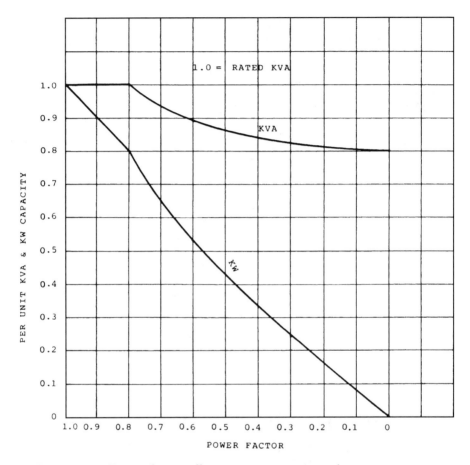

FIGURE 4.3 Power factor effect on generator capacity.

power factor or an adjustment to the kilowatt demand charge.

2. A high power factor reduces the load on transformers and distribution equipment.

3. A high power factor decreases the I^2R losses in transformers, distribution cable, and other equipment, resulting in a direct saving of kilowatt-hour power consumption.

4. A high power factor helps stabilize the system voltage.

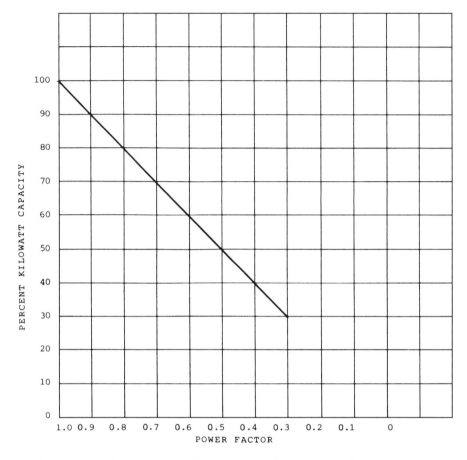

FIGURE 4.4 Power factor effect on transformer capacity.

4.4 HOW TO IMPROVE THE POWER FACTOR

To improve the power factor for a given load, the reactive load component (kvar) must be reduced. This component of reactive power lags the power component (kW input) by 90 electrical degrees, so that one way to reduce the effect of this component is to introduce a reactive power component that leads the power component by 90 electrical degrees. This can be accomplished by the use of a power capacitor, as illustrated in the power diagram in Fig. 4.5, resulting in a net decrease in the angle θ or an increase in the power factor.

Several methods are used to improve the power factor in a system installation. One method that can be employed in large systems is to use synchronous motors to drive low-speed loads that require continuous operation. A typical application for a synchronous motor is driving a low-speed air compressor, which provides process compressed air for the plant. The synchronous motor is adjusted to operate at a leading power factor and thus provide leading kilovars to offset the lagging kilovar of inductive-type loads such as induction motors.

Synchronous motors are usually designed to operate at an 80% leading power factor and to draw current that leads the line voltage rather than lags it, as is the case with induction motors and transformers. For example, consider a load of 2000 kW at a 70% lagging power factor. The utilization of a 200-hp synchronous motor operating at an 80% leading power factor will increase the overall system power factor from 70% lagging to 85% lagging.

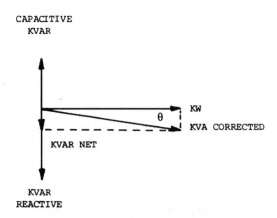

FIGURE 4.5 Vector diagram of power input with a power factor correction.

The more popular method of improving the power factor on low-voltage distribution systems is to use power capacitors to supply the leading reactive power required. The amount and location of the corrective capacitance must be determined from a survey of the distribution system and the source of the low–power factor loads. In addition, the total initial cost and payback time of the capacitor installation must be considered.

To reduce the system losses, the power factor correction capacitors should be electrically located as close to the low–power factor loads as possible. In some cases, the capacitors can be located at a particular power feeder. In other cases, with large-horsepower motors, the capacitors can be connected as close to the motor terminals as possible. The power factor capacitors are connected across the power lines in parallel with the low-power factor load.

The number of kilovars of capacitors required depends on the power factor without correction and the desired corrected value of the power factor.

The power factor and kilovars without correction can be determined by measuring the power factor, line amperes, and line voltage at the point of correction. For a three-phase system,

$$\text{kVA input} = \frac{\text{line amperes} \times \text{line volts} \times \sqrt{3}}{1000}$$

$$\text{kW} = \text{kVA} \times \text{PF}$$

$$\text{kvar} = \sqrt{(\text{kVA})^2 - (\text{kW})^2}$$

Or the line kilowatts, line amperes, and line voltage can be measured. Then,

$$\text{kVA input} = \frac{\text{line amperes} \times \text{line volts} \times \sqrt{3}}{1000}$$

$$\text{PF} = \frac{\text{kW}}{\text{kVA}}$$

$$\text{kvar} = \sqrt{(\text{kVA})^2 - (\text{kW})^2}$$

The capacitive kilovars required to raise the system to the desired power factor can be calculated as follows:

$$\text{kvar capacitance} = \text{kvar load} - \sqrt{\frac{1 - \text{PF}^2}{\text{PF}^2}\,(\text{kW load})^2}$$

where PF is the desired power factor.

For example, consider a 1000-kW load with a 60% power factor, which one wishes to correct to 90%:

kW load = 1000

$$\text{kvar load} = \sqrt{\left(\frac{1000}{0.60}\right)^2 - 1000^2} = 1333 \text{ kvar}$$

$$\text{kvar capacitance} = 1333 - \sqrt{\frac{1 - (0.90)^2}{(0.90)^2}\,1000^2} = 849$$

Tables such as Table 4.1 have been developed and are available from most power capacitor manufacturers to simplify this calculation. Table 4.1 provides a multiplier to be applied to the kilowatt load to determine the capacitive kilovars required to obtain the desired corrected power factor. Consider the same 1000-kW load with a 60% power factor which one wishes to correct to 90%. From Table 4.1, for the existing power factor (60%) and the corrected power factor (90%), the power factor correction factor is 0.849. Thus, the number of kilovars of capacitance required is 1000 × 0.849 = 849 kvar.

Let us verify this calculation:

$$\text{Uncorrected kVA} = \frac{1000}{0.60} = 1667$$

$$\text{Uncorrected lagging kvar} = \sqrt{1667^2 - 1000^2} = 1333$$

Correction capacitor kvar = 849

Net corrected lagging kvar = 484

TABLE 4.1 Power Factor Correction Factors

Existing power factor, %	Corrected power factor			
	100%	95%	90%	85%
60	1.333	1.005	0.849	0.714
62	1.265	0.937	0.781	0.646
64	1.201	0.872	0.716	0.581
66	1.138	0.810	0.654	0.519
68	1.078	0.750	0.594	0.459
70	1.020	0.692	0.536	0.400
72	0.964	0.635	0.480	0.344
74	0.909	0.580	0.425	0.289
76	0.855	0.526	0.371	0.235
78	0.802	0.474	0.318	0.183
80	0.750	0.421	0.266	0.130
81	0.724	0.395	0.240	0.104
82	0.698	0.369	0.214	0.078
83	0.672	0.343	0.188	0.052
84	0.646	0.317	0.162	0.026
85	0.620	0.291	0.135	
86	0.593	0.265	0.109	
87	0.567	0.238	0.082	
88	0.540	0.211	0.055	
89	0.512	0.184	0.028	
90	0.484	0.156		
91	0.456	0.127		
92	0.426	0.097		
93	0.395	0.067		
94	0.363	0.034		
95	0.329			

$$\text{Corrected kVA} = \sqrt{1000^2 + 484^2} = 1111$$

$$\text{Corrected power factor} = \frac{1000}{1111} = 0.90$$

Over the years, there have been several guidelines used for the selection of induction motor power factor correction capacitors. Three of these guidelines are as follows:

1. Add corrective kilovars of capacitors equal to 90% of the motor no-load kilovolt-amperes.
2. Add corrective kilovars of capacitors equal to 100% of the motor no-load kilovolt-amperes.
3. Add corrective kilovars of capacitors equal to 50% of the motor full-load kilovolt-amperes.

Table 4.2 is a comparison of these methods of selecting correction capacitors for some typical four-pole, 1800-rpm induction motors.

TABLE 4.2 Comparison of Power Factor Correction Methods

Motor rating, hp	Initial PF	90% NL kVA		100% NL kVA		50% FL kVA	
		kvar	Corr. PF	kvar	Corr. PF	kvar	Corr. PF
Standard NEMA design B, 1800-rpm induction motors							
10	0.843	3.70	0.979	4.12	0.987	5.14	0.999
25	0.853	8.70	0.980	9.66	0.988	12.45	1.000
50	0.863	12.37	0.962	13.74	0.970	23.64	1.000
100	0.903	19.50	0.973	21.67	0.978	45.20	0.997L
200	0.905	37.34	0.973	41.49	0.978	89.24	0.997L
Energy-Efficient, 1800-rpm induction motors							
10	0.850	2.98	0.967	3.31	0.976	4.92	0.999
25	0.867	7.24	0.977	8.05	0.984	11.71	1.000
50	0.805	18.18	0.962	20.20	0.974	24.86	0.993
100	0.861	24.16	0.960	26.84	0.968	46.85	1.000
200	0.897	51.75	0.974	45.81	0.980	87.44	0.998L

The question is: What is a typical motor in regard to power factor? Figure 4.6 shows the variation in full-load power factor for standard four-pole, 1800-rpm induction motors. Figure 4.6 is based on published data from 10 electric motor manufacturers. The difference in the full-load power factor for a specific horsepower rating can vary by 5 to 15 points. Therefore, it is best to know the power factor information on the specific motors requiring power factor correction. The no-load methods of selecting correction capacitors are conservative and increase the corrected power factor to 95% or higher. However, the no-load information is not readily available. In contrast, the full-load power factor and efficiency are generally available either as published literature or on the motor nameplate. These data can be used to calculate the motor power factor and input kilowatt-amperes. The use of the 50% full-load kilowatt-amperes to determine the corrective kilovars generally results in a

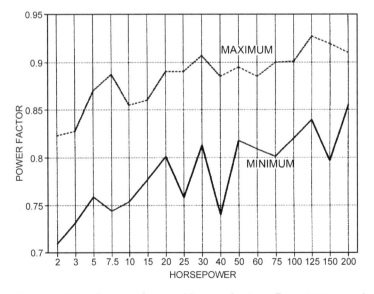

FIGURE **4.6** Power factor Nema design B, 1800-rpm induction motors.

corrected power factor of 0.99 or better to a slightly leading power factor. This method should be used with caution if the motor is not operating at full rated load. Under partial-load conditions, the corrected power factor can be over 0.90 leading. The higher the horsepower of the motor, the more likely it is that the corrected power factor can be leading at partial loads. A partial motor load is not an unusual condition. Studies indicate that the average load on induction motors rated 125 hp and larger ranges from 50 to 85% of full-load rating. For 1800-rpm induction motors the power factor at 50% load is usually 0.07 to 0.14 points lower than the power factor at full load. If the capacitor correction is not used to supply kilovars for other uncorrected motors on the same circuit, a value lower than 50% of the full-load input kilovolt-amperes should be used for the correction kilovars.

In the application of power factor correction capacitors at the motor location, NEMA recommends the following procedure based on the published or nameplate data for the electric motor:

1. The approximately full-load power factor can be calculated from published or nameplate data as follows:

$$PF = \frac{431 \times hp}{E \times I \times Eff}$$

where

> PF = per unit power factor at full load (per unit PF = percent PF/100)
> hp = rated horsepower
> E = rated voltage
> I = rated current
> Eff = per unit nominal full-load efficiency from published data or as marked on the motor nameplate (per unit Eff = percent Eff/100)

2. For safety reasons, it is generally better to improve the power factor for multiple loads as a part of the plant distribution system. In those cases in which local codes or other circumstances require improving the power

factor of an individual motor, the kilovar rating of the improvement capacitor can be calculated as follows:

$$\text{kvar} = \frac{0.746 \times \text{hp}}{\text{Eff}} \left(\frac{\sqrt{1 - \text{PF}^2}}{\text{PF}} - \frac{\sqrt{1 - \text{PF}_i^2}}{\text{PF}_i} \right)$$

where

kvar = rating of a three-phase power factor improvement capacitor

PF_i = improved per unit power factor for the
motor–capacitor combination

3. In some cases, it may be desirable to determine the resultant power factor PF_i, where the power factor improvement capacitor selected within the maximum safe value specified by the motor manufacturer is known. The resultant full-load power factor PF_i can be calculated from the following:

$$\text{PF}_i = \frac{1}{\sqrt{\{[(\sqrt{1 - \text{PF}^2})/\text{PF}] - [(\text{kvar} \times \text{Eff})/(0.746 \times \text{hp})]\}^2 + 1}}$$

Warning: In no case should power factor improvement capacitors be applied in ratings exceeding the maximum safe value specified by the motor manufacturer. Excessive improvement may cause overexcitation, resulting in high transient voltages, currents, and torques, which can increase safety hazards to personnel and cause possible damage to the motor or to the driven equipment. For additional information on safety considerations in the application of power factor improvement capacitors, see NEMA Publication No. MG2, *Safety Standard for Construction and Guide for Selection, Installation and Use of Electric Motors and Generators.*

The level to which the power factor should be improved depends on the economic payback in terms of the electric utility power factor penalty requirements and the system energy saved because of lower losses. In addition, the characteristic of the motor load must be considered. If the motor load is a cyclical load that varies from the rated load to a light load, the value of corrective kilovar capacitance should not result in a leading power factor at light loads.

To avoid this possibility, NEMA recommends that the maximum value of the corrective kilovars added be less than the motor no-load kilovar requirement by approximately 10%. Thus,

Maximum capacitor kvar for three-phase motors

$$= \frac{I_{NL} V \sqrt{3} \times 0.90}{1000}$$

$$= \frac{I_{NL} V}{642}$$

where

I_{NL} = motor no-load line current
V = motor line voltage

For example, consider a 50-hp, 1800-rpm induction motor operating on a 230-V, three-phase, 60-Hz power system. Table 4.3 shows the performance of this motor at various loads without power factor correction. Table 4.4 shows the full-load performance with various values of correction capacitor kilovars, including 100% of the no-load kilovolt-amperes (13.7 kvar) and 50% of the full-load

TABLE 4.3 Induction Motor Performance Without Power Factor Correction[a]

Load	Line	Eff.	PF	kW input	kVA	kvar
Full load	118.7	0.915	0.862	40.8	47.3	24.0
3/4 load	89.4	0.922	0.852	30.3	35.6	18.6
1/2 load	64.8	0.920	0.785	20.3	25.8	16.0
1/4 load	44.1	0.887	0.598	10.5	17.6	14.1
No load	34.4	0.000	0.073	1.0	13.7	13.7

[a] 50-hp, 1750-rpm, 230-V, three-phase, 60-Hz induction motor.

TABLE 4.4 Induction Motor and Capacitor Performance with Power Factor Correction at Full Load[a]

Capacitor kvar	System line current	System kVA	System PF	Line loss reduction, %
0	118.7	47.3	0.862	0
8	109.9	43.8	0.931	14
12	106.7	42.5	0.959	19
13.7	105.5	42.0	0.970	21
15	104.8	41.7	0.977	22
18	103.4	41.2	0.989	24
23.7	102.3	40.8	1.000	26

[a] 50-hp, 1750-rpm, 230-V, three-phase, 60-Hz induction motor.

kilovolt-amperes (23.7 kvar). These values of kilovars correct the power factor to 0.97 and unity, respectively. Based on 4000 hr/yr operation at 5¢/kWh for electric energy, a correction to unity power factor could result in a saving in energy costs of $70/yr. The combined motor–capacitor performance at partial loads is shown in Table 4.5. Note that at partial loads with the higher values of corrective kilovars the power factor can be very leading. Figure 4.7 shows the comparison of the corrected and uncorrected power

TABLE 4.5 Induction Motor and Capacitor Performance with Power Factor Correction at Various Loads[a]

Capacitor kvar	Net power factor				
	Full load	3/4 load	1/2 load	1/4 load	No load
0	0.862	0.852	0.785	0.598	0.073
8	0.931	0.944	0.930	0.865	0.174
12	0.959	0.977	0.981	0.981	0.516
13.7	0.97	0.987	0.994	0.999	0.999
15	0.977	0.993	0.999	(0.996L)	(0.599L)
18	0.989	1.000	(0.995L)	(0.937L)	(0.225L)
23.7	1.000	(0.986L)	(0.935L)	(0.738L)	(0.099L)

[a] 50-hp, 1750-rpm, 230 V, three-phase, 60-Hz induction motor.

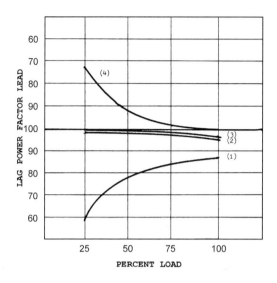

FIGURE 4.7 Power factor of 50-hp induction motor with various levels of kilovar correction: (1) no correction, (2) 12-kvar correction, (3) 13.7-kvar correction, (4) 23.7-kvar correction.

factor at various level of kilovar correction for the 50-hp induction motor. The high level of power factor correction should be avoided if the motor is going to be operating at partial loads and the capacitors are connected directly to the motor terminals. The application of capacitor kilovars up to the no-load kilowatt-amperes results in a lagging power factor for all load conditions.

The National Electric Code (NEC) has removed any restrictions on the size of power factor correction capacitors applied to induction motor circuits. This places the responsibility on plant electrical engineers to select the power factor correction strategies that best suit their plant operations.

4.4.1 Where to Locate Capacitors

The power factor correction capacitors should be connected as closely as possible to the low–power factor load. This is very often determined by the nature and diversity of the load. Figure 4.8 illustrates typical points of installation of capacitors:

At the Motor Terminals. Connecting the power capacitors to the motor terminals and switching the capacitors with the motor load is a very effective method for correcting the power factor. The benefits of this type of installation are the following: No extra switches or protective devices are required, and line losses are reduced from the point of connection back to the power source. Corrective capacitance is supplied only when the motor is operating. In addition, the correction capacitors can be sized based on the motor nameplate information, as previously discussed.

If the capacitors are connected on the motor side of the overloads, it will be necessary to change the overloads to retain proper overload protection of the motor. A word of caution: With certain types of electric motor applications, this method of installation can result in damage to the capacitors or motor or both.

Never connect the capacitors directly to the motor under any of the following conditions:

The motor is part of an adjustable-frequency drive system.
Solid-state starters are used.
Open transition starting is used.
The motor is subject to repetitive switching, jogging, inching, or plugging.
A multispeed motor is used.
A reversing motor is used.
There is a possibility that the load may drive the motor (such as a high-inertia load).

In all these cases, self-excitation voltages or peak transient currents can cause damage to the capacitor and motor. In these types of installations, the capacitors should be switched with a contactor interlocked with the motor starter.

At the Main Terminal for a Multimotor Machine. In the case of a machine or system with multiple motors, it is common practice to correct the entire machine at the entry circuit to the machine. Depending on the loading and duty cycle of the motors,

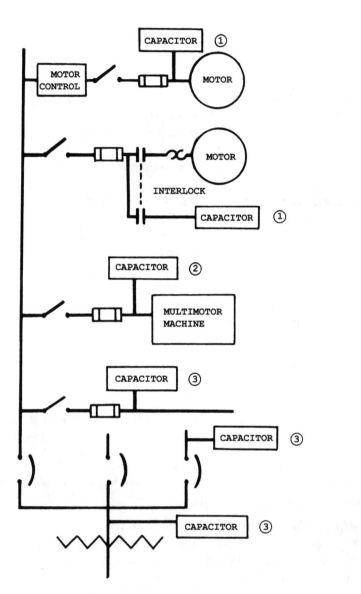

FIGURE 4.8 Where to install power factor capacitors.

it may be desirable to switch the capacitors with a contractor interlocked with the motor starters. In this manner, the capacitors are connected only when the main motors of a multimotor system are operating.

At the Distribution Center or Branch Feeder. The location of the capacitors at the distribution center or branch feeder is probably most practical when there is a diversity of small loads on the circuit that require power factor correction. However, again, the capacitors should be located as close to the low–power factor loads as possible in order to achieve the maximum benefit of the installation.

4.5 THE POWER FACTOR WITH NONLINEAR LOADS

The growing use of power semiconductors has increased the complexity of system power factor and its correction. These power semiconductors are used in equipment such as

> Rectifiers (converters)
> DC motor drive systems
> Adjustable-frequency AC drive systems
> Solid-state motor starters
> Electric heating
> Uninterruptible power supplies
> Computer power supplies

In the earlier discussion about the power factor in sinusoidal systems, only two components of power contributed to the total kilovolt-amperes and the resultant power factor: the active or real component, expressed in kilowatts, and the reactive component, expressed in kilovars. When nonlinear loads using power semiconductors are used in the power system, the total power factor is made up of three components:

1. *Active, or real, component*, expressed in kilowatts.
2. *Displacement component*, of the fundamental reactive elements, expressed in kilovars or kilovolt-amperes.

3. *Harmonic component.* The result of the harmonics and the distorted sinusoidal current and voltage waveforms generated when any type of power semiconductor is used in the power circuit, the harmonic component can be expressed in kilovars or kilovolt-amperes. The effect of these nonlinear loads on the distribution system depends on (1) the magnitude of the harmonics generated by these loads, (2) the percent of the total plant load that is generating harmonics, and (3) the ratio of the short-circuit current available to the nominal fundamental load current. Generally speaking, the higher the ratio of short-circuit current to nominal fundamental load current, the higher the acceptable level of harmonic distortion.

Therefore, more precise definitions of power factors are required for systems with nonlinear loads as follow:

Displacement power factor. The ratio of the active power of the fundamental in kilowatts to the apparent power of the fundamental in kilovolt-amperes.
Total power factor. The ratio of the active power of the fundamental in kilowatts to the total kilovolt-amperes.
Distortion factor, or *harmonic factor*. The ratio of the root-mean-square (rms) value of all the harmonics to the root-mean-square value of the fundamental. This factor can be calculated for both the voltage and current.

Figure 4.9 illustrates the condition in which the total power factor is lower than the displacement power factor as a result of the harmonic currents.

$$\text{Displacement power factor} = \frac{\text{kW}}{\sqrt{\text{kW}^2 + \text{fund. kvar}^2}}$$

$$\text{Total power factor} = \frac{\text{kW}}{\sqrt{\text{kW}^2 + (\text{fund. kvar} + \text{harm. kvar})^2}}$$

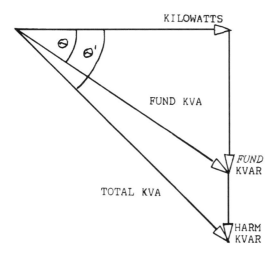

FIGURE 4.9 Power factor, nonsinusoidal system.

Unfortunately, conventional var-hour meters do not register the total reactive energy consumed by nonlinear loads. If the voltage is nonsinusoidal, the var-hour meter measures only the displacement volt-ampere-hours and ignores the distortion volt-ampere-hours. Therefore, for nonlinear loads, the calculated power factor based on kilowatt-hour and var-hour meter readings will be higher than the correct total power factor. The amount of the error in the power factor calculation depends on the magnitude of the total harmonic distortion.

The harmonics result from distorted AC line currents caused by the power semiconductor devices. Typical current wave shapes caused by AC adjustable-frequency drives are shown in Figs. 4.10 and 4.11. Figure 4.10 illustrates the wave shape of the AC line current produced by an adjustable-frequency drive system with the converter section containing silicon control rectifiers (SCRs) or other controllable power switching devices, such as those used in current source inverters and DC drive systems. The harmonic problem for this type of converter is complicated by the voltage notch and voltage spikes that occur during the switching of the converter solid-state devices. The displacement power factor for this type of

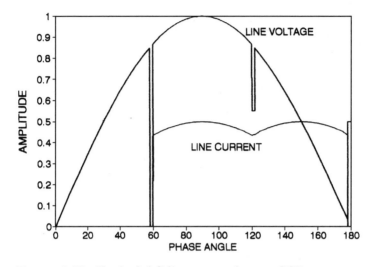

FIGURE 4.10   Typical AC line wave shapes, SCR converter.

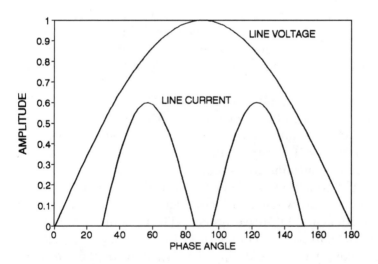

FIGURE 4.11   Typical AC line wave shapes, diode converter.

converter is linear with load. The total power factor and the harmonic component depend on the system reactance and short-circuit capacity. Figure 4.11 illustrates the wave shape of the AC line current produced by an adjustable-frequency drive system with the converter section operating as a voltage source with a typical diode bridge rectifier converter, such as those used in voltage source and pulse width modulation (PWM) inverters. Again, the total power factor and the current wave shape vary depending on the system impedance, the capacitance on the output of the converter, and the power semiconductor characteristics. The lower the line inductance, the higher the harmonics and the higher the value of the peak current. The displacement power factor for this type of converter is constant over the speed range. However, the total power factor depends on the harmonic distortion factor.

Figure 4.12 compares the displacement power factor for the SCR bridge converter and the diode bridge converter. For both types of converters, there is a difference between the displacement

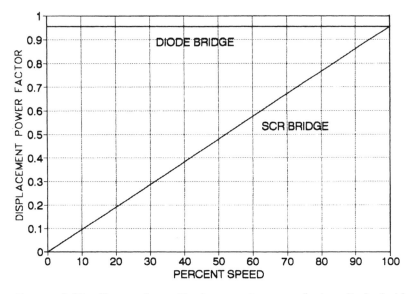

FIGURE 4.12 Comparison displacement power factor diode bridge versus SCR bridge.

power factor and the total power factor. Again, the total power factor depends on the harmonics generated, the harmonic distortion factor, and the power system characteristics.

4.6 HARMONICS AND THE POWER FACTOR

How do you know if you have a harmonic distortion problem?

A single adjustable-frequency drive or DC drive is usually not a problem. However, if over 20% of the plant load contains power semiconductors, there is a good possibility of problems. Some of these problems are

Transformer overheating and noisy
Electric motors overheating and noisy
Power factor correction capacitors overheating or failing
Unexplained tripping of circuit devices
Computer malfunctions
Low system power factor

The application of power factor correction capacitors without an analysis of the system can aggravate rather than correct the problem, particularly if the fifth and seventh harmonics are present.

Since it is not economical to eliminate the harmonics from individual drives or devices, an analysis should be made of the total system to determine the major harmonics present and the methods for reducing the harmonics to acceptable levels. Only then can the proper harmonic filters or traps and power factor correction capacitors be applied at the best location in the system.

The American National Standard, ANSI/IEEE 519–1981 IEEE Guide for Harmonic Control and Reactive Compensation of Static Power Converters, is a guide for making a system analysis. This guide discusses the harmonics generated, the AC line voltage notches, and the harmonic distortion factor. The standard recommends the maximum harmonic current distortion based on the ratio of maximum short-circuit current to maximum load current and suggests locations of harmonic filters. Table 4.4 of the above standard suggests voltage distortion limits for medium- and high-voltage power systems, and Fig. 27 of the standard shows the theoretical voltage distortion versus short-circuit ratio for six pulse rectifiers. In addition, there are a number of articles in the technical literature that discuss various procedures for determining harmonic distortion and power factor correction, with examples of systems that have been corrected.

A wide variety of instrumentation is available to perform the necessary harmonic analysis of power system or specific load. A few of these are listed below.

- *Harmonimeter*, manufactured by Myron Zucker Inc., Royal Oak, Michigan. This instrument is shown in Fig. 4.13. The unit is battery operated and is quite portable. It is equipped with a clamp-on current transformer, and it measures current harmonics from the second to nineteenth harmonic as a percent of total current. Data in this harmonic range are adequate in many cases. The instrument is easy to use to locate the source of the high-harmonic loads since measurements can be made wherever a clamp-on current transformer can be installed.

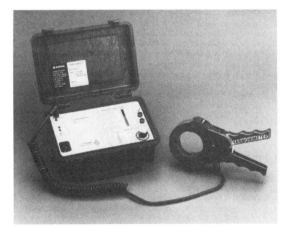

FIGURE 4.13 Myron Zucker Harmonimeter. (Courtesy Myron Zucker Inc., Royal Oak, MI.)

- *BMI 3030/3060 Power Profiler*, manufactured by Basic Measuring Instruments, Foster City, California. This instrument is reasonably portable. The unit can measure rms current, voltage, true power, apparent power, true power factor, displacement power factor, and total harmonic distortion. In addition, with the harmonic option package, it can provide a graphic presentation of the voltage or current and an analysis of the harmonics as a percent of the fundamental up to the fiftieth harmonic. Figure 4.14 is a photograph of this instrument. This unit can provide a printout of the voltage and current wave shapes as well as the harmonic spectrum, giving magnitude and phase of each harmonic up to the fiftieth harmonic.
- *Dranetz Disturbance Waveform Analyzer*, Series 656A, manufactured by Dranetz Technologies, Inc., Edison, New Jersey. This unit is also portable. The unit has a cathode display screen and a thermal printer for data output. The unit can measure the harmonic distortion for voltage and current, total harmonic distortion as a percent of the fundamental, and individual harmonics

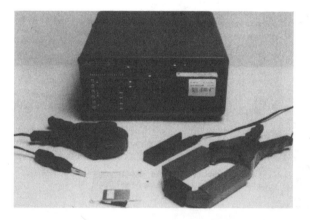

FIGURE 4.14 BMI 3030/3060 Power Profiler. (Courtesy Basic Measuring Instruments, Foster City, CA.)

and phase shift of each harmonic up to the fiftieth. Figure 4.15 is a photograph of this unit. The following figures illustrate the output capability of this unit. Figure 4.16 is a graphic printout of the phase current for a DC motor drive. Figure 4.17 is the printout of the total harmonic distortion, the odd harmonic contribution, and the even harmonic contribution. In addition, the printout shows the percent and phase angle of each harmonic through the fiftieth. Note that in this example, the major harmonics are the fifth at 33.8% and the eleventh at 8.6%.

4.6.1 System Example

How does this apply to a typical case of a power system with high harmonic content? Consider the case of an industrial plant in which the major circuit has a line current wave shape as shown in Fig. 4.18. This circuit had a mixture of DC drives, adjustable-frequency drives, and standard induction motors. The total current harmonic distortion (THD) was measured at 19.67%, the fifth harmonic at 16.89%, and the voltage harmonic distortion at 4.13%. For comparison, the equivalent rms sine-wave current has been superimposed on Fig. 4.18. After an analysis of the harmonic content of the system, correction equipment was added particularly to reduce the

FIGURE 4.15 Dranetz series 656A Disturbance Waveform Analyzer. (Courtesy of Dranetz Technologies, Inc., Edison, NJ.)

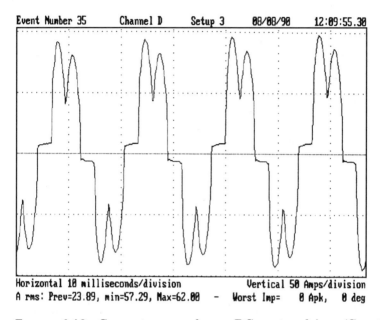

Event Number 35   Channel D   Setup 3   00/00/90   12:09:55.30

Horizontal 10 milliseconds/division   Vertical 50 Amps/division
A rms: Prev=23.89, min=57.29, Max=62.00 - Worst Imp= 0 Apk, 0 deg

FIGURE 4.16 Current wave shape, DC motor drive. (Courtesy of Dranetz Technologies, Inc., Edison, NJ.)

fifth harmonic. Figure 4.19 shows the line current wave shape after the system correction. Although this doesn't look like a perfect sine wave, the total current harmonic distortion was reduced to 7.92%, the fifth harmonic to 2.79%, and the voltage harmonic distortion to 1.25%. Figure 4.20 is a comparison of the harmonic content before and after correction. The total harmonic current distortion factor of 7.92% is acceptable on many power systems, but the eleventh harmonic of 5.6% may be borderline. If further improvement is required, it would involve reducing the eleventh harmonic.

4.7 POWER FACTOR MOTOR CONTROLLERS

In recent years, solid-state control devices have been developed that, when connected between a power source and an electric motor,

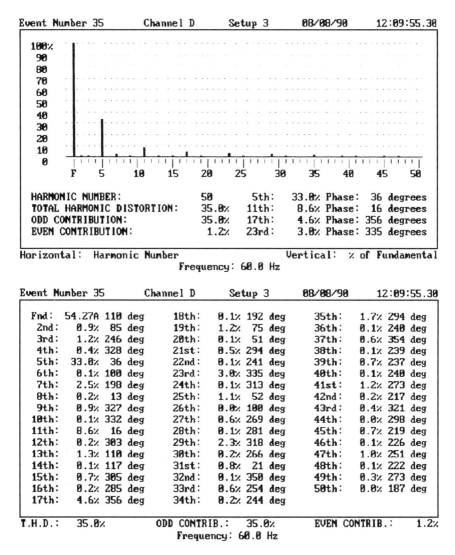

FIGURE 4.17 Graphic and harmonic analysis of current of a DC motor drive. (Courtesy of Dranetz Technologies, Inc., Edison, NJ.)

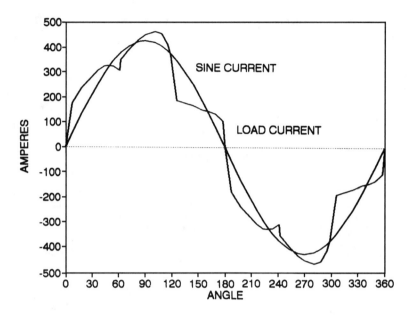

FIGURE 4.18 System line current before harmonic suppression.

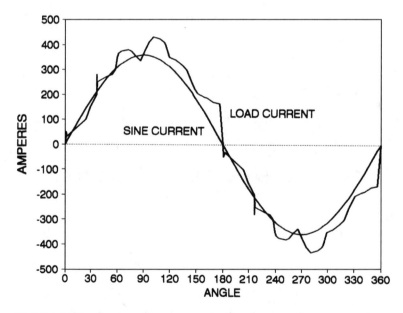

FIGURE 4.19 System line current after harmonic suppression.

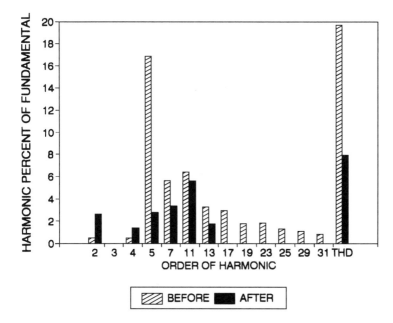

FIGURE 4.20 System harmonic current comparison before and after harmonic suppression.

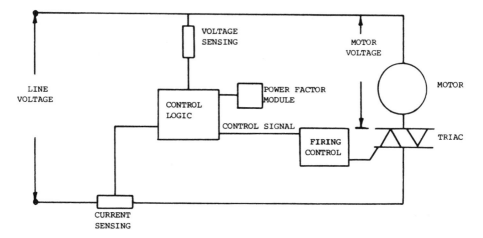

FIGURE 4.21 Single-phase power factor controller block diagram.

maintain an approximately constant power factor on the motor side of the controller. These devices are generally called *power factor controllers*. Most of the units are made under a license of U.S. Patent 4,052,648 issued to F. J. Nola and assigned to NASA.

The controller varies the average voltage applied to the motor as a function of the motor load and thus decreases the motor losses at light-load requirements.

4.7.1 Single-Phase Motors

For application to single-phase motors, the power factor controller consists of a triac, sensing and control circuits, and a firing circuit for the triac, as shown in Fig. 4.21. The power factor controller sensing circuit monitors the phase angle between the voltage and current and produces a signal proportional to the phase angle. This signal is compared to a reference signal that indicates the desired phase angle. This comparison produces an error signal that provides the timing for firing the triac or SCR and causes the phase angle to remain constant when the load changes. Typical motor voltage and current waveforms are shown in Figs. 4.22 and 4.23.

If the phase angle increases, the control circuit adjusts the triac firing angle to decrease the average voltage applied to the motor. Conversely, if the phase angle decreases, the control circuit adjusts the firing angle of the triac to increase the average voltage applied to the motor.

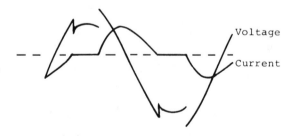

FIGURE 4.22 Single-phase power factor controller with no load on the motor.

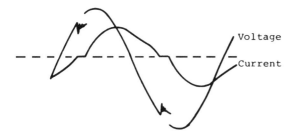

FIGURE 4.23 Single-phase power factor controller with a full load on the motor.

The power factor of the motor is the cosine of the phase angle between the motor voltage and current. Therefore, with this control system, by maintaining the phase angle constant, the motor operates at an approximately constant power factor over the load range. The maximum power factor is the power factor of the motor at the rated load with the triac full on. The minimum power factor will be determined by the minimum voltage setting for no-load operation. This voltage setting must be high enough to provide stable operation and prevent the motor from stalling on the sudden application of load. However, the lower the no-load voltage, the higher the power savings at no load.

How are power savings achieved by decreasing the motor voltage at light loads? The motor losses can be grouped into three categories:

1. Constant losses, such as friction and windage
2. Magnetic core losses, which are some function of the applied voltage
3. I^2R losses, which are a function of the square of the motor current, including rotor losses

For a given load condition, the net losses, and hence the motor power input, decrease with a decrease in voltage as long as the magnetic core losses decrease more than the I^2R losses increase. In addition, there is some increase in losses due to harmonics added to the motor input voltage by the triac switching and the losses in the controller.

In some instances, the increased harmonic content of the input voltage will result in increased motor noise.

The amount of power saved with a power factor controller depends on the duty cycle of the application. Typical power savings under various loads and duty cycles are shown in Fig. 4.24. The power savings are shown as a percent of the full voltage input and as a function of the percent running times at full load versus running at a light load. To result in significant power savings, at least 50% of

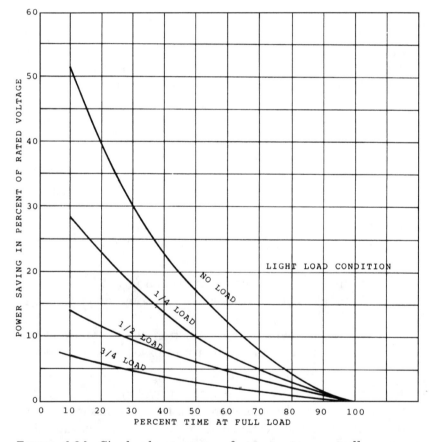

FIGURE 4.24 Single-phase power factor motor controller power savings.

the running time should be at one-fourth load or less. Typical applications of this type may be drill presses and cutoff saws used in production processes.

Figure 4.22 shows an oscilloscope picture of the motor voltage and current at no load for a single motor controlled by a power factor controller.

Figure 4.23 shows an oscilloscope picture of the motor voltage and current of the same motor with load applied to the motor. Note the constant angle between the zero crossing of the voltage and current in both cases.

4.7.2 Three-Phase Motors

More recently, the application of power factor motor controllers has been extended to three-phase motors. In some cases, this has been

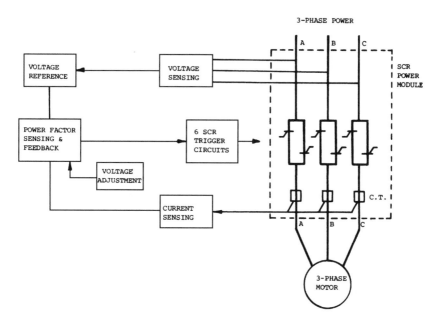

FIGURE 4.25 Three-phase power factor motor controller block diagram.

accomplished by adding a power-saver module to existing solid-state three-phase motor controllers. These solid-state controllers generally include other features such as current limit, timed acceleration, phase unbalance, undervoltage, and overload protection.

The power factor control function is accomplished by sensing the phase angle between the motor voltage and current. This signal is fed back and compared with a reference, and the difference is used to feed the input signal voltage to the six SCRs in the power module.

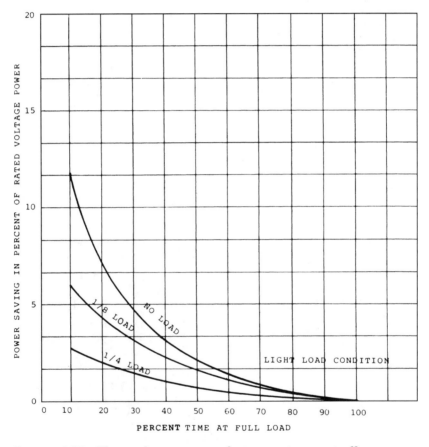

FIGURE 4.26 Three-phase power factor motor controller power savings.

The feedback voltage from the power factor sensing circuit will change the average voltage applied to the motor in accordance with the load on the motor. This reduces both the motor current and voltage under light-load conditions. The circuit is designed to react to load changes to prevent stalling of the motor on instantaneous load changes. Most of the controllers have provisions for setting the minimum no-load voltage; this voltage is generally 65% of rated full voltage. Figure 4.25 is a typical block diagram for the three-phase controller.

The three-phase power factor controllers have potential applications in which the duty cycle for the motor is varying from light or no load to full load as a step function. Examples of potential applications are ripsaws, conveyors, rock crushers, and centrifuges.

The potential power saving when a power factor controller is applied to a three-phase motor is substantially lower than when such a controller is applied to a single-phase motor. Figure 4.26 illustrates the power saving when the controller is applied to a three-phase motor for various duty cycles and loads. These curves depend on the ratio of the no-load losses of the motor. However, it appears that the power factor controller shows significant power savings only on those three-phase motor applications in which the motor operates at no load or light loads over 75% of the operating time.

To apply a power factor controller properly, the load characteristics, motor characteristics, and load cycle must be known. In addition, one must determine how the controller–motor combination will respond to the load cycle. Only then can the potential power saving and economic payback analysis be made.

5

Applications of Induction Motors

5.1 GENERAL DISCUSSION

The AC induction motor is used more than any other means to power industrial equipment. This is confirmed by the U.S. Department of Energy report on electric motors, which states that 53–58% of the electric energy generated is consumed by electric motors (see Table 5.1). Because there are so many applications, it is impossible to develop a list or guide for all the applications of AC induction motors.

However, a guide for the selection of three-phase induction motors for many applications had been developed and is shown in Table 5.2. The table outlines the matching of the driven load requirements to the electric motor characteristics. In applying this guide, it must be recognized that the purpose of the horsepower ratings and NEMA standards for electric motors is to define the useful performance range of the motors in a way most intelligible to the user for fixed-frequency applications. For variable-frequency applications, see Sec. 5.6 on motor selection for adjustable-fre-

TABLE 5.1 Electric Motor Population and Energy Consumption, 1977

Horsepower	No. of motors (000s)	New sales, average for 1973–1977 (000s)	Annual electric energy consumption, billions of kWh
1–5	54,583	3567	34
5.1–20	10,421	573	103
21–50	3,313	151	155
51–125	1,703	59	338
126 and over	1,004	35	573
Total	71,024	4385	1203

Source: U.S. Department of Energy Report DOE/CS-0147, 1980.

quency power supplies. The fixed-frequency rating of induction motors includes six major variables:

Supply voltage and frequency
Number of phases
Rated horsepower
Torque characteristics
Speed
Temperature rise

In addition, the basis of rating specifies the type of duty:

Continuous duty
Intermittent duty
Varying duty

It is desirable to use standard motors for as many different applications as possible. Consequently, general-purpose continuous-rated motors should be used when

1. The peak momentary overloads do not exceed 75% of the breakdown torque
2. The root-mean-square (rms) value of the motor losses over an extended period of time does not exceed the losses at the service factor rating

TABLE 5.2 Three-Phase Electric Motor Selection Chart

For this type of equipment	Requiring these torques		With these load characteristics	Type and description[a]
	Starting	Max. running		
Water supply pumps Industrial and chemical pumps Cooling towers Air-handling equipment Compressors Conveyors Process machinery Petroleum and chemical process equipment	100–150% of full-load torque	200–250% of full-load torque	Continuous operation, constant speed, high speed (over 720 rpm), easy starting; subject to short time overloads; good speed regulation	*Energy-efficient:* NEMA design B, normal torques: normal starting current; can be used with variable-frequency/variable-voltage inverters; higher efficiency than standard design B motors
Centrifugal pumps Blowers and fans Drilling machines Grinders Lathes Compressors Conveyors	100–150% of full-load torque	200–250% of full-load torque	Variable load conditions, constant speed; subject to short time overloads; good speed regulation	*NEMA design B,* normal torques: normal starting current; can be used with variable-frequency/variable-voltage inverters

Applications	Starting torque		Characteristics of load	Motor type
Reciprocating pumps Stokers Compressors Crushers Ball and rod mills	200–300% of full-load torque	Not more than full-load torque	High starting torque due to high inertia, back pressure, standstill friction, or similar mechanical conditions; torque requirements decrease during acceleration to full-load torque; not subject to severe overloads; good speed regulation	*NEMA design C*, high torque: normal starting current; not recommended for use with variable-frequency inverters
Punch presses Cranes Hoists Press brakes Shears Oil well pumps Centrifugals	Up to 300% of full-load torque	200–300% of full-load torque; loss of speed during peak loads required	Intermittent loads; may require frequent start, stop, and reverse cycles; machine uses a flywheel to carry peak loads; poor speed regulation to smooth power peaks; may require acceleration of high-inertia load	*NEMA design D*, high torque: high slip; standard types have slip characteristics of 5–8% or 8–13% slip

TABLE 5.2 Continued

For this type of equipment	Requiring these torques		With these load characteristics	Type and description[a]
	Starting	Max. running		
Blowers Fans Machine tools Mixing machines Conveyors Pumps	Some require low torque; others require several times full-load torque	200% of full-load torque at each speed	Speed selection is desired, and two, three, or four fixed speeds are sufficient; starting torque can be low on blowers to high on conveyors; metal-cutting machines are usually constant hp; friction loads (conveyors) are usually constant torque; fluid or air loads (blowers) are variable torque	*Multispeed:* general normal torque on dominant winding or speed; consequent pole windings or separate windings for each speed; based on load requirement, can be constant horsepower, constant torque, variable torque

Crushers Conveyors Bending rolls Ball and rod mills Centrifugal blowers Pumps Printing presses Cranes and hoists Centrifugalsl	Can provide torque up to maximum torque at standstill	200–300% of full-load torque	Loads that require very high starting torque with low starting current; require speed adjustment over limited range (2 to 1); torque control during acceleration or controlled acceleration	*Wound-rotor*: requires rotor control system to provide desired characteristic; control may be resistors or fixed-frequency inverters in the secondary (rotor) circuit; actual load speed depends on setting of rotor control

ᵃ See Chapter 1 for a detailed description.

3. The duration of any overload does not raise the momentary peak temperature above a value safe for the motor's insulation system

5.1.1 Energy-Efficient Motors

The selection of an energy-efficient motor should be based on several additional factors:

1. Electric power-saving and life-cycle cost comparison to standard motors
2. Improved ability to perform under adverse conditions such as abnormal voltage (See Secs. 5.3 and 5.4 for performance comparisons to standard motors. Note the superior performance of energy-efficient motors under abnormal voltage conditions.)
3. Lower operating temperatures
4. Noise level
5. Ability to accelerate higher-inertia loads than standards motors
6. Higher operating efficiencies at all load points. (Figure 5.1 illustrates this comparison on a 25-hp, 1765-rpm polyphase induction motor. Note that at all loads the energy-efficient motor presents an opportunity for energy savings.)

In general, energy-efficient motors can be justified on a payback basis because of the annual saving of electric energy. This saving is a function of the hours of operation per year and kilowatt-energy reduction. For example, consider a 25-hp, 1800-rpm application with an average annual operating time of 4000 hr and a cost of electric power of 5¢/kWh:

Standard motor efficiency = 88%

$$\text{Power input} = \frac{25 \times 0.746}{0.88} = 21.19 \text{ kW}$$

Energy−efficient motor efficiency = 93.0%

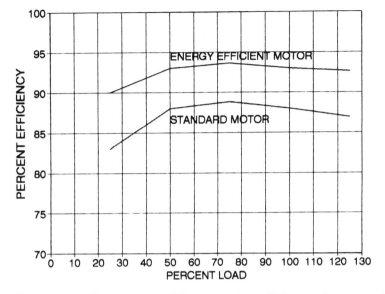

FIGURE 5.1 Comparison of the operating efficiency of energy-efficient versus standard polyphase induction motors at 25 hp and 1765 rpm.

$$\text{Power input} = \frac{25 \times 0.746}{0.93} = 20.05\,\text{kW}$$

$$\text{Annual energy saving} = (21.19 - 20.05) \times 4000$$

$$= 4560\,\text{kWh}$$

$$\text{Annual power cost saving} = 4560 \times 0.05 = \$228$$

Typical list price standard drip–proof motor = \$993

Typical list price energy–efficient, drip–proof motor

$$= \$1226$$

$$\text{Time to recover initial cost} = \frac{1226 - 993}{228} = 12\,\text{months}$$

thus indicating a very favorable cost/benefit ratio for this application.

This method of cost/benefit analysis is approximate but is generally acceptable if the time to recover the initial investment is less than 3 yr. However, when it is desirable, more accurate methods can be used that consider the increasing cost of power, the required return on the investment, and the product useful life. The detailed procedures for making this type of economic analysis are described in Chapter 7.

In certain applications and duty cycles, energy-efficient motors cannot be justified on the basis of energy saved; for example:

1. Intermittent-duty or special torque applications:

    Hoists and cranes
    Traction drives
    Punch presses
    Machine tools
    Oil field pumps
    Fire pumps
    Centrifugals

2. Types of loads:

    Multispeed
    Frequent starts and stops
    Very high-inertia loads
    Low-speed motors (below 720 rpm)

Additional factors that should be considered in the selection and applications of electric motors are reviewed in the following sections of this chapter.

5.2 VARYING DUTY APPLICATIONS

In many applications, the load imposed on the driving motor varies from no load to a peak load. When the motor load fluctuates, the temperature rise of the motor fluctuates. When there is a definite repeated load cycle, the motor size selection can be based on the rms value of motor losses for the load cycle. However, normally, the losses at each increment of the load cycle are not available to the user. Therefore, a good approximation for the motor size selection can be based on the rms horsepower for the load cycle. The rms horsepower is then defined as that equivalent steady-state horsepower that would result in the same temperature rise as that of the defined load cycle. When making the rms calculation, it is assumed that, when the motor is running, the heat dissipation is 100% effective. However, when the motor is at standstill, the heat dissipation is severely reduced and is limited to dissipation by radiation and natural convection. This can be compensated for by using an effective cooling time at standstill of one-fourth of the total standstill time. *An important word of caution*: This method of selecting electric motors is not satisfactory for applications requiring frequent starting or plug reversing or systems with a high load inertia.

5.2.1 Sample Calculation

Duty cycle	
40 hp	15 min
20 hp	20 min
10 hp	5 min
Stop	5 min
Total cycle	45 min

$$
\begin{aligned}
\text{hp}^2 \times t = 40^2 \times 15 \quad &= 24{,}000 \\
20^2 \times 20 \quad &= 8{,}000 \\
10^2 \times 5 \quad &= 500 \\
0 \times 5 \quad &= 0 \\
\text{hp}^2 \times t \text{ total} &= 32{,}500
\end{aligned}
$$

$$
T_e = \text{effective cooling time} = 15 + 20 + 5 + 1/4 \times 5
$$
$$
= 41.25 \text{ min}
$$

$$
\text{rms hp} = \sqrt{\frac{\text{hp}^2 \times t}{T_e}} = \sqrt{\frac{32{,}500}{41.25}} = 28 \text{ hp}
$$

From a thermal standpoint, a 30-hp standard motor would be satisfactory for this application.

Is the ratio of peak horsepower to nameplate (NP) horsepower satisfactory?

$$\frac{\text{Peak hp}}{\text{NP hp}} = \frac{40}{30} \times 100 = 133\%$$

Based on a limit of 150% for the ratio of peak horsepower to motor nameplate horsepower, the 30-hp motor could be satisfactory for this load.

Consider a slightly different cycle:

Duty cycle	
40 hp	10 min
20 hp	25 min
10 hp	10 min
Total cycle	45 min

$$40^2 \times 10 \quad = 16{,}000$$

$$20^2 \times 25 \quad = 10{,}000$$

$$10^2 \times 10 \quad = 1{,}000$$

$$\text{hp}^2 \times \text{total} = 27{,}000$$

Effective cooling time $= 10 + 25 + 10 = 45$ min

$$\text{rms hp} = \sqrt{\frac{27{,}500}{45}} = 24.5 \text{ hp}$$

From a thermal standpoint, a standard 25-hp motor would be satisfactory. However,

$$\frac{\text{Peak hp}}{\text{NP hp}} = \frac{40}{25} \times 100 = 160\%$$

Based on a limit of 150% for this ratio, the use of a 25-hp motor is not considered satisfactory, and a 30-hp motor should be used.

5.3 VOLTAGE VARIATION

NEMA Standard MG1 recognizes the effect of voltage and frequency variation on electric motor performance. The standard recommends that the voltage deviation from the motor rated voltage not exceed $\pm 10\%$ at the rated frequency. A certain degree of confusion may exist in regard to the rated motor voltage since the rated motor voltage and the system voltage are different. The rated motor voltage has been selected to match the utilization voltage available at the motor terminals. This voltage allows for the voltage drop in the power distribution system and for voltage variation as the system load changes.

The basis of the NEMA standard rated motor voltages for three-phase, 60-Hz induction motors is as follows:

System voltage	Rated motor voltage
208	200
240	230
480	460
600	575

For single-phase, 60-Hz induction motors, the basis for standard rated motor voltages is as follows:

System voltage	Rated motor voltage
120	115
240	230

Polyphase induction motors are designed to operate most effectively at their nameplate rated voltage. Most motors will operate satisfactorily over $\pm 10\%$ voltage variation, but deviations from the nominal motor design voltage can have marked effects on the motor

TABLE 5.3 Effect of Voltage Variation on Polyphase Induction Motor Performance

Operating characteristic	Effect of voltage change		
	90% voltage	110% voltage	120% voltage
Starting and maximum running			
torque	Decrease 19%	Increase 21%	Increase 44%
Synchronous speed	No change	No change	No change
Percent slip	Increase 23%	Decrease 17%	Decrease 30%
Full-load speed	Decrease 1%	Increase 1%	Increase 1%
Starting current	Decrease 10–12%	Increase 10–12%	Increase 25%
Magnetic noise, any load	Decrease slightly	Increase slightly	Noticeable increase
Standard NEMA design B motors			
Efficiency			
Full load	Increase 0.5–1%	Decrease 1–4%	Decrease 7–10%
3/4 load	Increase 1–2%	Decrease 2–5%	Decrease 6–12%
1/2 load	Increase 2–4%	Decrease 4–7%	Decrease 14–18%
Power factor			
Full load	Increase 8–10%	Decrease 10–15%	Decrease 10–30%

3/4 load	Increase 10–12%	Decrease 10–15%	Decrease 10–30%
1/2 load	Increase 10–15%	Decrease 10–15%	Decrease 15–40%
Full-load current	Increase 1–5%	Increase 2–11%	Increase 15–35%
Temperature rise at full load	Increase 6–12%	Increase 4–23%	Increase 30–80%
Energy-efficient NEMA B motors			
Efficiency			
Full load	Decrease 1–2%	Increase 0.5–1%	Small increase
3/4 load	Minimal change	Minimal change	Decrease 0.5–2%
1/2 load	Increase 1–2%	Decrease 1–2%	Decrease 7–20%
Power factor			
Full load	Increase 1%	Decrease 3%	Decrease 5–15%
3/4 load	Increase 2–3%	Decrease 4%	Decrease 10–30%
1/2 load	Increase 4–5%	Decrease 5–6%	Decrease 10–30%
Full-load current	Increase 11%	Decrease 7%	Decrease 11%
Temperature rise at full load	Increase 23%	Decrease 14%	Decrease 21%

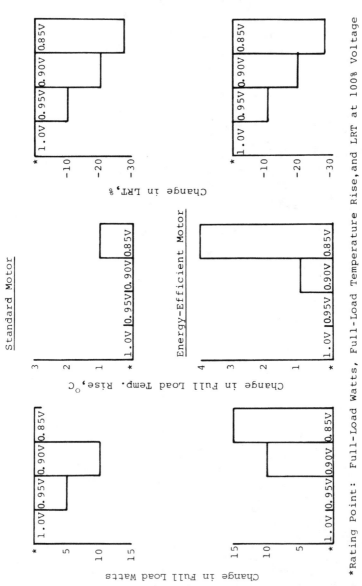

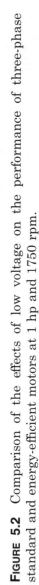

*Rating Point: Full-Load Watts, Full-Load Temperature Rise, and LRT at 100% Voltage

FIGURE 5.2 Comparison of the effects of low voltage on the performance of three-phase standard and energy-efficient motors at 1 hp and 1750 rpm.

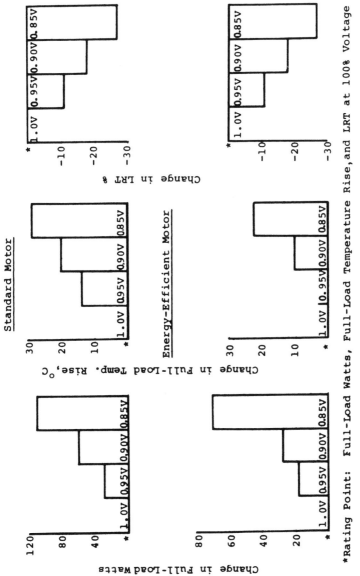
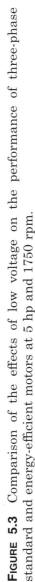

FIGURE 5.3 Comparison of the effects of low voltage on the performance of three-phase standard and energy-efficient motors at 5 hp and 1750 rpm.

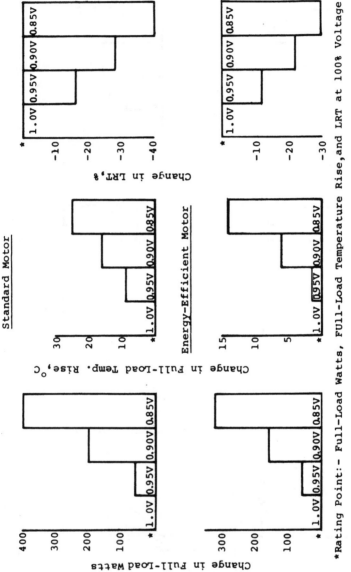
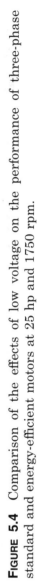

FIGURE 5.4 Comparison of the effects of low voltage on the performance of three-phase standard and energy-efficient motors at 25 hp and 1750 rpm.

performance. Table 5.3 indicates the type of changes in performance to expect with variation in the motor terminal voltage. The table shows the effect on the efficiency and power factor in standard NEMA design B motors and also in energy-efficient motors. It is important to note that the efficiency and the power factor of energy-efficient motors are not as sensitive to voltage variations as standard motors.

In recent years, the trend in some areas is to decrease system voltage to reduce the system load. In some cases, this reduction has been as low as 85% of the nominal voltage. For most electric motor loads, this increases rather than decreases the electric motor input and increases the full-load temperature rise. Also, the locked-rotor torque is severely reduced such that hard-to-start loads may not start at the 85% voltage level. Figures 5.2–5.4 illustrate the effect of reduced voltage on selected horsepower ratings of both standard motors and energy-efficient motors.

5.4 VOLTAGE UNBALANCE

Voltage unbalance can be more detrimental than voltage variation to motor performance and motor life. When the line voltages applied to a polyphase induction motor are not equal in magnitude and phase angle, unbalanced currents in the stator windings will result. A small percentage voltage unbalance will produce a much larger percentage current unbalance.

Some of the causes of voltage unbalance are the following:

1. An open circuit in the primary distribution system.
2. A combination of single-phase and three-phase loads on the same distribution system, with the single-phase loads unequally distributed.
3. An open wye-delta system.

   a. Variation in ground supply impedance: An increase in primary ground impedance increases the voltage and current balances. Maximum unbalance occurs with overloaded transformers, and the large single-phase load is in the lagging phase. The motor serves to balance the system voltage better when the motor is loaded than when it is unloaded.

b. Transformer loading varied 50 to 150%: The greatest unbalance occurs when a smaller transformer is lightly loaded and a larger transformer is overloaded. If a single-phase load varies over a large range, it is better to supply this phase with the larger transformer on the leading phase.

c. Impedance of lines to the single-phase loads: The voltage and current unbalance ratios increase with the line impedances. Again, the unbalance ratios decrease as the motor is loaded more heavily.

d. Impedance of the supply line to the motor: The voltage and current unbalance ratios decrease with an increase in the line impedance to the motor. However, this results in lower voltage at the motor and decreased motor torque and speed.

e. Other parameters: Variations in the magnitude of transformer impedances, the power factor of single-phase loads, and primary line impedances have minor effects (not more than 3%) on the phase currents and unbalance ratios.

4. An open delta-delta system: When the two transformers are supplied by three-phase conductors, the only difference is in the lack of neutral impedance. Therefore, under usual conditions, the open delta-delta configuration will show superior performance to the open wye-delta configuration. However, when there are unequal line impedances or unusually long supply lines, there are additional observations.

a. There are mixed effects with variation of the lines supplying the single-phase loads.

b. An increase in the common primary supply line impedance results in increased voltage and current unbalances.

The unbalanced line voltages introduce negative sequence voltages in the polyphase motor. This negative sequence voltage produces an air gap flux rotating in a direction opposite to the rotor, thus producing high currents in the motor. A small negative sequence voltage can produce motor currents considerably in excess of those present under balanced voltage conditions.

NEMA Standard MG1 defines the percent voltage unbalance as follows:

$$\text{Percent voltage unbalance} = 100 \times \frac{\text{maximum voltage deviation from average voltage}}{\text{average voltage}}$$

These unbalanced voltages will result in unbalanced currents on the order of 6 to 10 times the voltage unbalance. Consequently, the temperature rise of the motor operating at a particular load and voltage unbalance will be greater than for the motor operating under the same conditions with balanced voltages. In addition, the large unbalance of the motor currents will result in nonuniform temperatures in the motor windings. An example of the effect of unbalanced voltages on performance is illustrated in Table 5.4 for a 5-hp motor.

Voltages should be evenly balanced as closely as possible. *Operation of a motor above 5% voltage unbalance is not recommended.* Even at 5% voltage unbalance, motor current unbalance on the order of 40% can exist.

In recognizing the detrimental effect of unbalanced line voltage on electric motor performance, NEMA Standard MG1 recommends derating motors that are applied to unbalanced systems, in accordance with Fig. 5.5 (NEMA MG1-14.35):

> When the derating factor is applied, the selection and setting of the overload device should take into account the combination of the derating factor applied to the motor and the increase in current resulting from the unbalanced voltages. This is a complex problem involving the variation in motor current as a function of load and voltage unbalance in addition to the characteristics of the overload device relative to I_{maximum} or I_{average}. In the absence

TABLE 5.4 Effect of Voltage Unbalance on Motor Performance[a]

Characteristic	Performance		
Average voltage	230	230	230
Percent unbalanced voltage	0.3	2.3	5.4
Percent unbalanced current	0.4	17.7	40.0
Increased temperature rise, °C	0	30	40

[a] 5-hp, 1725-rpm, 230-V, three-phase, 60-Hz motor.

of specific information, it is recommended that overload devices be selected and/or adjusted at the minimum value that does not result in tripping for the derating factor and voltage unbalance that applies. When unbalanced voltages are anticipated, it is recommended that the overload devices be selected so as to be responsive to $I_{maximum}$ in preference to overload devices responsive to $I_{average}$.*

The order of magnitude of the current unbalance is influenced not only by the system voltage unbalance but also by the system impedance, the nature of the loads causing the unbalance, and the operating load on the motor. Figure 5.6 indicates the range of unbalanced currents for various motor load conditions and system voltage unbalance.

The effect on other electric motor characteristics can be summarized as follows:

1. *Torques.* The locked-rotor and breakdown torques are decreased. If the voltage unbalance should be extremely severe, the torques might not be adequate for the application.
2. *Full-Load Speed.* The full-load speed is reduced slightly.

* Reprinted by permission from NEMA Standard Publication No. MG1-1987, *Motors and Generators*, copyright 1987 by the National Electrical Manufacturers Association.

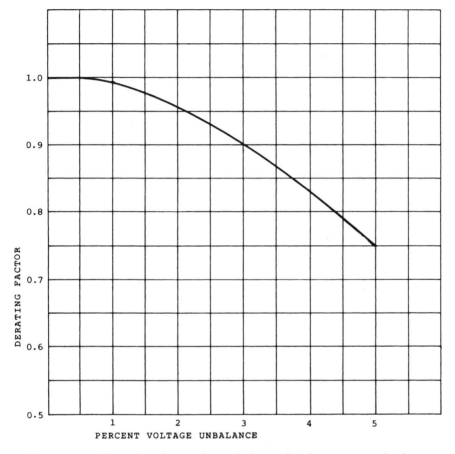

FIGURE 5.5 Derating factor for unbalanced voltages on polyphase induction motors. (Reprinted by permission from NEMA Standards Publication No. MG1-1987, *Motors and Generators*, copyright 1987 by the National Electrical Manufacturers Association.)

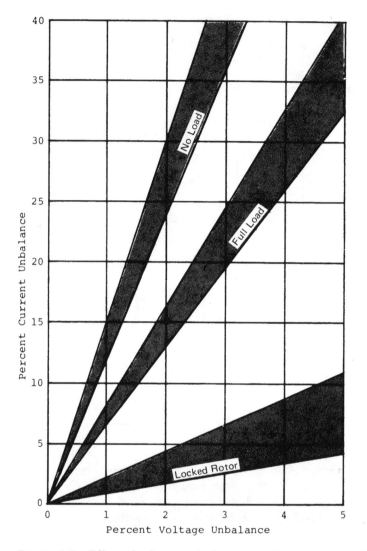

FIGURE 5.6 Effect of voltage unbalance on polyphase induction motor currents.

3. *Locked-Rotor Current.* The locked-rotor current will be unbalanced to the same degree that the voltages are unbalanced, but the locked-rotor kilovolt-amperes will increase only slightly.
4. *Noise and Vibration.* The unbalanced voltages can cause an increase in noise and vibration. Vibration can be particularly severe on 3600-rpm motors.

5.5 OVERMOTORING

In many instances, the practice has been to overmotor an application, i.e., to select a higher-horsepower motor than necessary. The disadvantages of this practice are

Lower efficiency
Lower power factor
Higher motor cost
Higher controller cost
Higher installation costs

One example of overmotoring is illustrated by the case of the varying duty applications discussed in Sec. 5.2. Consider the comparisons of the 40-hp motor that could have been selected based on the peak load versus the 30-hp motor that can be selected on the basis of the duty cycle:

1. Motor cost: list price of standard open 1800-rpm drip-proof motor:

   30 hp = $1160
   40 hp = $1446

2. Control Cost: NEMA-1 general-purpose motor, 240-V starter:

   30 hp, size 3 = $600
   40 hp, size 4 = $1350

This results in a cost difference of $1036, or 59%.

Figure 5.7 shows the difference in the input watts and Fig. 5.8 the difference in the input kilovolt-amperes for 30- and 40-hp motors

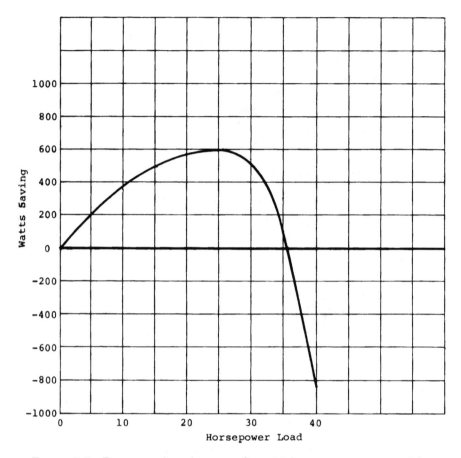

FIGURE 5.7 Power savings in watts for a 30-hp motor versus a 40-hp motor at the same load.

operating at the same output. At loads above 36 hp, the input is more favorable for the 40-hp motor. However, at loads below 36 hp, the kilowatt and kilovolt-ampere inputs are lower with the 30-hp motor.

In general, the larger the difference between the actual load and the motor rating, the higher the input requirements for the same load.

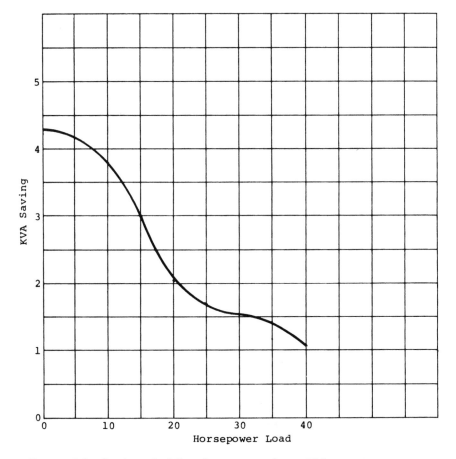

FIGURE 5.8 Savings in kilovolt-amperes for a 30-hp motor versus a 40-hp motor at the same load.

5.6 POLYPHASE INDUCTION MOTORS SUPPLIED BY ADJUSTABLE-FREQUENCY POWER SUPPLIES

When applying polyphase induction motors to adjustable-frequency power supplies, it must be remembered that the induction motor nameplate rating is based on a fixed-frequency, fixed-voltage, sine-

wave voltage source. Therefore, the application of polyphase induction motors with adjustable-frequency power supplies requires
consideration of additional features that are not considered for
fixed-frequency motor applications.

5.6.1 Types and Characteristics of Loads

Most electric motor loads fall in one of the following categories:

> Variable-torque loads
> Constant-torque loads
> Constant-horsepower loads
> Cyclical, intermittent, or varying loads

Variable-Torque Loads. For this type of load, the torque
usually increases as the square of the speed, and the horsepower
increases as the cube of the speed. This type of load is typical of fans
and centrifugal pumps within their normal operating range. The
characteristic of a variable-torque load is illustrated in Fig. 5.9. The

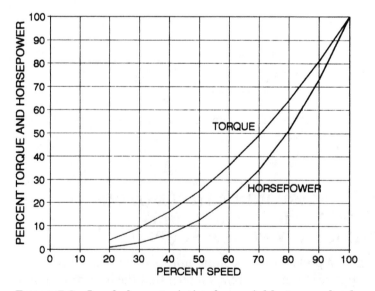

FIGURE 5.9 Load characteristics for variable-torque loads.

selection of the induction motor horsepower rating when supplied by a non-sine-wave power source for a variable-torque load can be based on the following:

> Standard motor, derated to 85 or 90% of nameplate horsepower rating
> Standard motor with 1.15 service factor, no derating
> Energy-efficient motor, no derating

Since the horsepower required by the variable-torque load increases as the cube of the driven speed, the motor will be overloaded if it is operated above its base speed.

Constant-Torque Loads. The motor torque required for this type of load is constant over the operating range and is not a function of speed. The motor horsepower required is proportional to the output speed. Typical constant-torque loads are conveyors, cranes, positive-displacement pumps, and mixers. The characteristic of this type of load is illustrated in Fig. 5.10. The selection of the

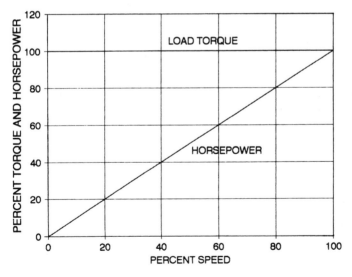

FIGURE 5.10 Load characteristics for constant-torque loads.

motor horsepower rating for constant-torque loads depends on the speed range of continuous operation.

1. For 2-to-1 speed range (i.e., 60–30 Hz) continuous operation:

    a. Standard motors derated 85–90% of nameplate horsepower
    b. Standard motors, class F insulation, 1.15 service factor, no derating
    c. Energy-efficient motors, no derating

2. For 6-to-1 speed range continuous operation (i.e., 60–10 Hz):

    a. Standard motors, increase up to two sizes over the base horsepower required.
    b. Standard motors, class F insulation, 1.15 service factor, increase up to one size over base horsepower required.
    c. Energy-efficient motors, increase up to one size over base horsepower required.
    d. For operation below 10 Hz, special cooling for the motor usually required.
    e. For more specific information, consult the motor manufacturer and the inverter manufacturer.

Constant-Horsepower Loads. For a constant-horsepower load, the load torque decreases as the speed increases, and the horsepower required remains essentially constant. This type of load is typical of metal-cutting machinery operating at a constant cutting or grinding velocity. As the diameter of the work is decreased, the drive speed is increased to maintain constant cutting velocity. The operating range is above base speed (i.e., 60 Hz and higher); therefore, the induction motor base speed must be selected so that at the maximum operating speed, the safe operating speeds of the induction motor and the driven equipment are not exceeded. The motor manufacturer and the equipment manufacturer should be consulted to determine the maximum safe speed for the system. These types of drives generally operate at a fixed voltage above base speed rather than constant volts/hertz. It must be determined that the induction motor will develop the horsepower required over the speed range required. Figure 5.11 shows the operating capability of a 10-hp standard induction motor in the constant-horsepower range (i.e., 60–120 Hz), and Fig. 5.12 illustrates the load characteristic for a constant-horsepower load.

Cyclical and Intermittent Loads. These types of loads require special consideration in sizing both the induction motor and the adjustable-frequency power supply. The motor manufacturer and the adjustable-frequency power supply manufacturer should be consulted for these types of applications.

5.6.2 Characteristics of an Induction Motor Operating on an Adjustable-Frequency Power Supply

Induction Motor Torques. The polyphase induction motor has a unique speed-torque curve at each frequency when operated at constant volts/hertz. This family of speed-torque curves is illus-

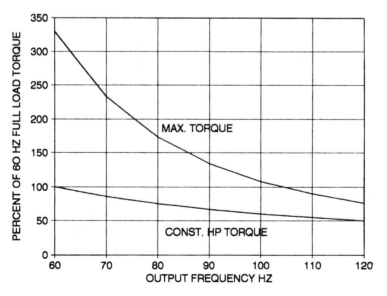

FIGURE 5.11 Performance of a 10-hp induction motor in the constant-horsepower range.

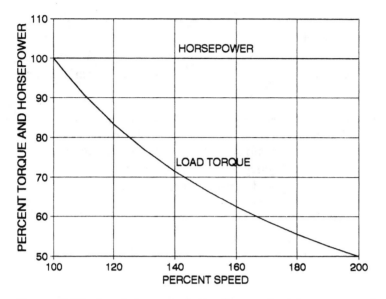

FIGURE 5.12 Load characteristics for constant-horsepower loads.

trated in Fig. 5.13 for a 100-hp, 1800-rpm, energy-efficient induction motor. The utilization of the motor torque capability depends on the current restriction imposed by the adjustable-frequency power supply. In most cases, the power inverter has a limit of 150% of its current rating for 1 min. Figure 5.14 illustrates the maximum torque (breakdown torque) capabilities of a 10-hp, four-pole induction motor under various voltage and current conditions. The maximum torque at 150% of motor rated current requires voltage boosts below 10 Hz, with the percent of voltage boost increasing as the frequency is decreased. The motor and adjustable-frequency power supply cannot operate continuously at these values of torques and current. These torques are important, however, when determining the acceleration time for high-inertia loads since the 150% current can be maintained only by the adjustable-frequency power supply for about 60 sec.

The locked-rotor torque or breakaway torque is a function of both the voltage and frequency applied to the motor. The locked-rotor torque per ampere increases as the frequency decreases, and the voltage is at constant volts/hertz down to about 6 Hz. The stator

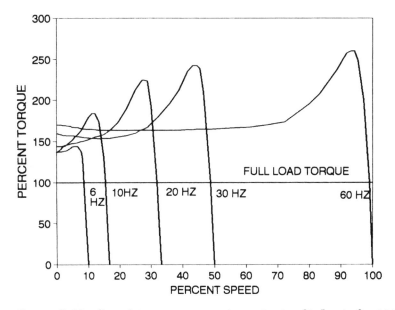

FIGURE 5.13 Speed-torque curves at constant volts/hertz for 100-hp, 1800-rpm, energy-efficient induction motor.

resistance may limit the torque that the motor develops at 6 Hz and lower. Figure 5.15 illustrates the low-frequency locked-rotor performance for a 10-hp, four-pole induction motor as a function of frequency and applied voltage. Locked-torque values at locked current equal to full-load current are shown by curve 2, Fig. 5.15. The voltages to achieve these values of torque are shown by curve 3, Fig. 5.15. This represents considerable voltage boost below about 3 Hz. Curve 4, Fig. 5.15, illustrates the voltage boost required to obtain locked torque equal to full-load torque. This indicates that locked torque equal to full-load torque can be obtained at rated voltage at about 6 Hz.

The actual value of locked-rotor torque or breakaway torque available depends on the motor characteristics and on the voltage boost capability of the adjustable-frequency power supply.

The starting torque and acceleration torque requirements may be a consideration, depending on the type of load and the inertia of the connected load. For most pumps and fans, this is usually not a

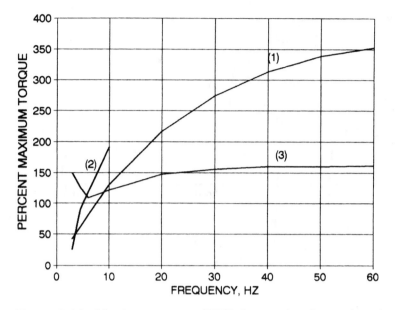

FIGURE 5.14 Maximum torque (BDT) for a 10-hp, four-pole induction motor. (1) Maximum torque (BDT) at constant volts/hertz. (2) Maximum torque (BDT) with 20% voltage boost. (3) Maximum torque (BDT) at 150% rated current (with voltage boost below 10 Hz).

problem. However, for a large fan with high inertia or other types of loads with high inertia such as a large flywheel, the time to accelerate the load becomes important. This time to accelerate the load can be expressed as

$$t = \frac{WK^2 \times RPM}{308 \times T}$$

where

t = time to accelerate, sec
WK^2 = moment of inertia of load referred to the motor plus the motor inertia, lb-ft^2
RPM = change in speed, i.e., low speed to high speed
T = net accelerating torque

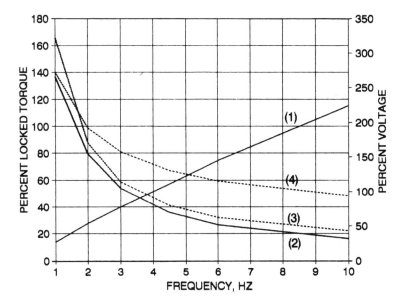

FIGURE 5.15 Locked-rotor performance at low frequencies for a 10-hp, four-pole, 60-Hz induction motor. (1) Percent locked torque at constant volts/hertz. (2) Percent locked torque at locked current equal to full-load current. (3) Percent voltage for locked current equal to full-load voltage. (4) Percent voltage for locked torque equal to full-load torque.

The net accelerating torque is the difference between the motor torque available and the torque required to drive the load. The motor torque available is a function of the voltage and current limitations imposed on the motor by the adjustable-frequency power supply. Therefore, the combination of the motor and adjustable-frequency power supply must be selected so that the load can be accelerated within the current–time constraint of the power supply and the acceleration time required for the application.

Induction Motor Efficiency. The efficiency of polyphase induction motors at constant-torque loads decreases as the frequency is decreased with constant-volt/hertz input. Figure 5.16 illustrates the efficiency of a 100-hp, 1800-rpm standard induction motor at var-

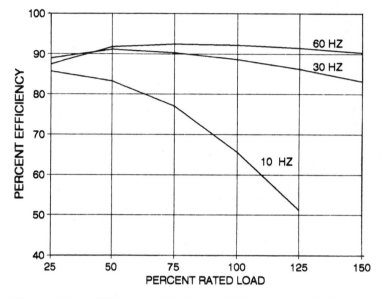

FIGURE 5.16 Efficiency of 100-hp, 1800-rpm standard motor at constant volts/hertz, sine power.

ious loads and frequencies when operated from a sine-wave power source. Similarly, Fig. 5.17 illustrates the performance of a 100-hp, 1800-rpm energy-efficient induction motor at various loads and frequencies when operated from a sine-wave power source.

Note that the energy-efficient motor has higher efficiency at all loads and frequencies and that this difference becomes more significant as the frequency is decreased. When the power source to the induction motor is nonsinusoidal and contains harmonics, additional losses, particularly stator and rotor winding losses and stray losses, are generated in the induction motor. The magnitude of the increased losses depends on the harmonic frequencies and voltages in the power source and also on the induction motor design.

Generally, the energy-efficient types of polyphase induction motors have a smaller increase in harmonic losses than the standard induction motors because of their lower stator and rotor resistances. Induction motors with deep-bar or T-bar rotors have higher har-

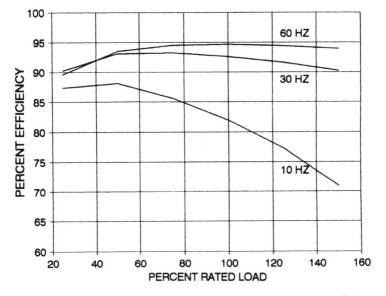

FIGURE 5.17 Efficiency of 100-hp, 1800-rpm energy-efficient motor at constant volts/hertz, sine power.

monic losses than induction motors with shallow-bar rotors. Induction motors with double-cage rotor construction can have excessive harmonic losses.

Figures 5.18 and 5.19 illustrate the increases in losses or decrease in efficiency for a particular motor/adjustable-power supply combination. There are no typical values for the increase in losses due to the harmonics since the increase depends on the magnitude and frequency of the harmonics generated by the adjustable-frequency power supply and the motor reaction to these harmonics. The increase in losses can range from as low as 10% to over 50%. Figure 5.20 shows a comparison of the efficiencies of a 100-hp, 1800-rpm standard motor and a 100-hp energy-efficient motor when supplied by a non-sine power source. Note the superior performance of the energy-efficient motor.

The six-step type of adjustable-frequency power supply generally causes a larger increase in induction motor losses than the pulse width modulation adjustable-frequency power supply.

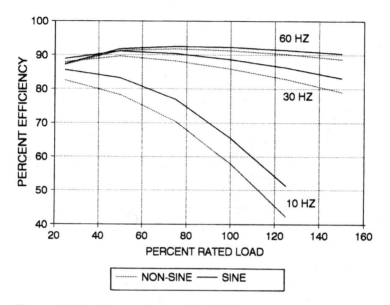

FIGURE 5.18 Comparison of a 100-hp, 1800-rpm standard motor efficiency with a sine-wave and a non-sine-wave power source.

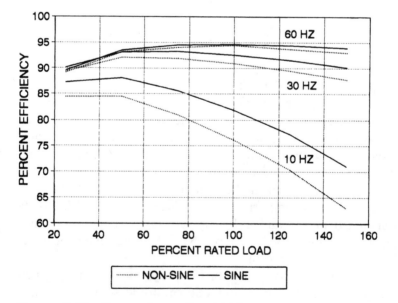

FIGURE 5.19 Comparison of a 100-hp, 1800-rpm energy-efficient motor efficiency with a sine-wave and a non-sine-wave power source.

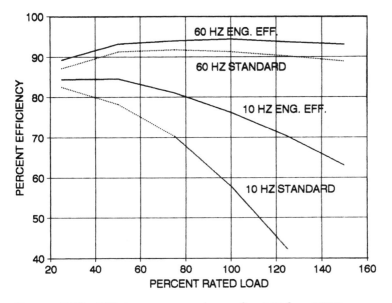

FIGURE 5.20 Efficiency comparison of a 100-hp, 1800-rpm energy-efficient motor versus a 100-hp, 1800-rpm standard motor, both with non-sine power.

Induction Motor Thermal Capacity. The induction motor nameplate identifies the type of motor enclosure, class of insulation, and service factor that the motor is rated for. These features determine the thermal capability of the motor. However, this rating is based on a sine power supply at rated voltage and frequency.

The heat dissipation of an induction motor depends on the class of insulation used in the motor, allowable temperature rise, ventilation, and type of enclosure. This varies from one motor manufacturer to another. However, in general, the allowable heat dissipation relative to the base-speed (60-Hz) heat dissipation decreases with speed (frequency) as shown in Fig. 5.21. Note that the heat-dissipation ability of the totally enclosed fan-cooled (TEFC) motor does not decrease as fast as that of the open motor. This is due to the ability of the TEFC to dissipate more of the heat generated by radiation and natural convection when operating at low speeds. The final temperature rise of the motor depends on heat-producing losses that must be dissipated, which are a function of the load,

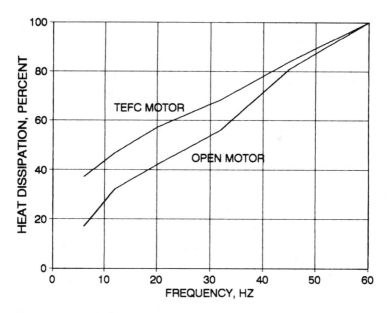

FIGURE 5.21 Relative heat-dissipation ability as a function of motor frequency (speed).

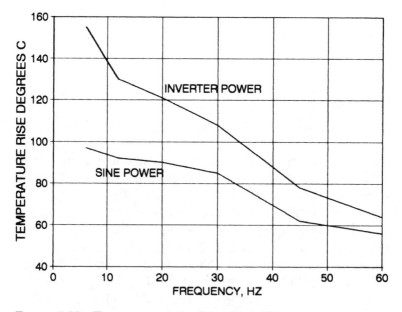

FIGURE 5.22 Temperature rise for a 5-hp, 1800-rpm open motor operating on a sine-wave power source versus an inverter power source.

speed, and harmonic losses caused by the motor input voltage. Again, the six-step adjustable-frequency power supply produces higher harmonic losses in the motor than the pulse width modulation power supply and results in higher motor temperatures.

The operation of a 5-hp, open, 1800-rpm induction motor on sine power supply and on an inverter with non-sine power supply are compared in Fig. 5.22. This figure compares the temperature rise for the same output torque at each frequency. The load was constant down to 30 Hz and was reduced at 20 Hz and lower to obtain constant temperature rise on the tests with the inverter power supply.

5.6.3 Summary: Induction Motor Selection for Adjustable-Frequency Inverter Systems

1. Define the load characteristics:

   a. Starting or breakaway torque required
   b. Type of load, i.e., variable torque, constant torque, constant horsepower, or cyclical
   c. Base horsepower
   d. Speed range for continuous operation
   e. Acceleration time and load inertia

2. Define the type of adjustable-frequency inverter.
3. Determine the motor horsepower rating and derating based on the torque requirements, speed range, and thermal capacity, considering the constraints imposed on the motor by the adjustable-frequency inverter.

6

Adjustable-Speed Drives and Their Applications

As environmental and other concerns slow the growth of electrical energy generation in coming years, it becomes essential that we conserve and use limited and precious resources more efficiently. Conserving electricity and making it a better energy source relies on the widespread adoption of the power conversion process, which takes electricity from a source and converts it to a form exactly suited to the electrical load.

Electric motors consume more than 60% of all electrical power in the United States. Adjustable-speed drives (ASDs) can improve the efficiency of these motors by about 50% in many applications. They can also reduce costs considerably. Power electronics allows us to develop efficient speed and torque control of electric motors at low costs. This, in turn, calls for development of optimized electro-mechanical power conversion units.

Today's technology requires different speeds in many areas where electric machines are used. Electric machines that use traditional control methods have mainly two states—stop and operate at maximum speed. Adjusting speed in these machines is costly and hardware dependent. This is because an increase in the machine size and mechanical parts requires more maintenance and less efficiency. After the discovery of semiconductors and the introduction of semiconductor devices such as diodes and transistors to industry, ASDs have become very popular because of their advantages over traditional control methods. The machine size is smaller and requires less maintenance and hardware.

Using ASDs, the speed of a motor or generator (electric machine) can be controlled and adjusted to any desired speed. Besides adjusting the speed of an electric machine, ASDs can also keep an electric machine speed at a constant level where the load is variable. For example, if the desired speed of the conveyer showed in Fig. 6.1 is 1 m/sec at any time, changing the load will not change the speed (Fig. 6.2).

The variable load can be controlled by just reducing the speed of the electric machine, which means applying less power to the system. As stated above, traditional control methods adjust the speed of the machine to the maximum level, and the machine works

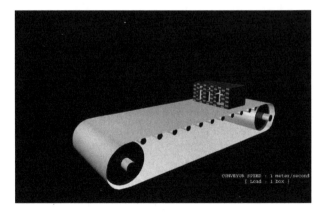

FIGURE 6.1 Typical conveyer maintaining a constant speed of 1 m/sec with one box as the load.

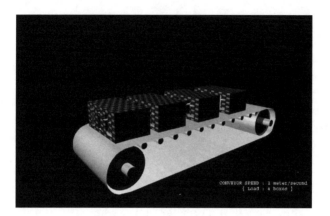

FIGURE 6.2   Typical conveyer maintaining a constant speed of 1 m/sec with four boxes as the load.

at that level regardless of the load. In addition, if the load increases, the speed decreases. However, adjustable speed drives can adjust the speed of the machine when the load changes, thereby reducing the applied power to the system.

6.1   THE IMPORTANCE OF ELECTRIC MOTOR DRIVES

Electric motors impact almost every aspect of modern living. Refrigerators, vacuum cleaners, air conditioners, fans, computer hard drives, automatic car windows, and multitudes of other appliances and devices all use electric motors to convert electrical energy into useful mechanical energy. In addition to running the commonplace appliances that we use every day, electric motors are also responsible for a very large portion of industrial processes. Electric motors are used at some point in the manufacturing process of nearly every conceivable product that is produced in modern factories. Because of the nearly unlimited number of applications for electric motors, it is not hard to imagine that there are over 700 million motors of various sizes in operation across the world. This enormous number of motors and motor drives has a significant impact on the world because of the amount of power they consume.

The systems that controlled electric motors in the past suffered from very poor performance and were very inefficient and expensive. In recent decades, the demand for greater performance and precision in electric motors, combined with the development of better solid-state electronics and cheap microprocessors has led to the creation of modern ASDs. An ASD is a system that includes an electric motor as well as the system that drives and controls it. Any adjustable speed drive can be viewed as five separate parts: the power supply, the power electronic converter, the electric motor, the controller, and the mechanical load.

The power supply is the source of electric energy for the system. The power supply can provide electric energy in the form of AC or DC at any voltage level. The power electronic converter provides the interface between the power supply and the motor. Because of this interface, nearly any type of power supply can be used with nearly any type of electric motor. The controller is the circuit responsible for controlling the motor output. This is accomplished by manipulating the operation of the power electronic converter to adjust the frequency, voltage, or current sent to the motor. The controller can be relatively simple or as complex as a microprocessor. The electric motor is usually, but not always, a DC motor or an AC induction motor. The mechanical load is the mechanical system that requires the energy from the motor drive. The mechanical load can be the blades of a fan, the compressor of an air conditioner, the rollers in a conveyor belt, or nearly anything that can be driven by the cyclical motion of a rotating shaft.

Electric motor drive technology is constantly evolving and expanding to new applications. More advanced electric motor drives are now replacing older motor drives to gain better performance, efficiency, and precision. Advanced electric motor drives are capable of better precision because they use more sophisticated microprocessor or DSP controllers to monitor and regulate motor output. They also offer better efficiency by using more efficient converter topologies and more efficient electric motors. The more advanced drives of today also offer a performance boost by utilizing superior switching schemes to provide more output power while using lighter motors and more compact electronics.

6.2 MOTOR DRIVE PARAMETERS

There are several criteria by which an electric motor drive can be evaluated. The main criteria are efficiency, power factor, harmonic distortion, size, cost, and power density ratio.

Efficiency is one of the most important criteria of modern electric motor drives. Efficiency is simply the mechanical power delivered to the mechanical load divided by the total electrical power consumed by the motor drive. Efficiency is expressed as a percent. For example, if an electric motor drive has an efficiency rating of 75%, exactly three-quarters of the electricity consumed by the motor drive is converted into useful mechanical energy. The remaining one-quarter is lost in the form of heat in the electronics and in the motor. Efficiency is obviously of great importance because of the vast numbers of motor drives throughout the world.

The power factor is technically defined as the cosine of the angle between the voltage and the current supplied to the motor drive. If the AC voltage and current supplied to a motor drive are visualized as sine waves, the power factor quantitatively represents how close the two sine waves are to lining up. If the sine waves of the voltage and current perfectly line up, the power factor is unity. If the sine waves are completely opposite of each other, the power factor is zero. Higher power factors (as close to unity as possible) are desired because they reduce the losses in the electrical power system. Electric utility companies charge an extra fee if the power factor of an industrial load is not above a minimum value. Low power factors cause losses in the power system and cause power quality problems.

Harmonic distortion can occur when a power electronic converter in a motor drive draws a nonsinusoidal current from the power system. There are several other sources of harmonic distortion, including high-intensity discharge lighting, power electronic power supplies, etc. Harmonic distortion can cause serious adverse effects on other equipment running on the same electrical system.

Basically, harmonic distortion is a power quality concern that affects the electrical system and other equipment running on the electrical system.

Size and cost are also very critical in evaluating an electric motor drive. The size/weight of a motor drive ultimately determines the feasibility of a motor drive for a particular application. The cost of a motor drive is obviously an important concern in most situations; however, the importance of cost may be overestimated in most cases. In most applications, the cost of the motor/motor drive makes a very small percentage of the money that will be spent on the motor. The vast majority of the expense in most applications is the cost of the energy to run the motor throughout its life. For example, a large industrial motor may cost $5000 to purchase and install, but will cost $70,000 in electricity costs to keep it running for its 10-yr life span. This is one major reason that efficiency is so important. An energy-efficient motor drive may cost significantly more than a conventional motor drive, but the capital cost is usually quite small in comparison with the energy costs. In most cases, a more expensive energy-efficient motor drive more than recovers its larger initial cost. This is a very important concept that is often overlooked.

The power/density ratio is important in many applications where space is limited. The power/density ratio is the ratio of power output of a motor drive to the weight or size of the motor drive. Power/density ratio is particularly crucial in vehicular applications, i.e., automotive and aerospace applications, where size and weight are constrained.

Each of the parameters of motor drives described has a different level of importance. The size, cost, and power density ratio of a motor drive determine its suitability for a given application. Each of these has little impact on anyone but the end user of the motor drive. The power factor and the harmonic distortion of a given motor drive are more important properties because they determine the effects on power quality. Power quality issues potentially affect more than just the end user. Motor drive efficiency is the most important property of motor drives in general because efficiency affects everyone. The end user pays for inefficient motor drives in the form of higher electricity costs and society pays for the energy losses in the form of economic waste and ecological damage.

6.3 THE IMPACT OF MOTOR EFFICIENCY

The efficiencies of motors and motor drive systems used throughout the world have a large impact on the world in a number of ways. The energy consumed by electric motors accounts for nearly half of all electricity generated in the world today. Because motors make up such a large portion of the world's electrical load, the inefficiency of today's motors results in enormous amounts of wasted energy. The impact of wasted energy in the form of electric motor losses can be broken down into two main categories: economic and ecological. With the constant increase in electricity costs, energy lost due to inefficient motors translates directly to wasted money. Since approximately 75% of the power generated in the United States comes from the burning of fossil fuels, wasted energy due to motor inefficiency needlessly increases the amount of pollution created in the power generation process.

There are currently more than 700 million motors in service worldwide, and approximately 50 million new motors are manufactured every year. These motors come in many different sizes and are used for countless different applications. In the United States (and most other industrialized nations) motors can be classified very roughly into two categories: industrial and residential. Industrial motors are used in applications ranging from mining to manufacturing to the climate control of large commercial buildings. Industrial motors are generally integral horsepower motors, ranging in size from one to several thousand horsepower. Residential motors are used in applications ranging from air conditioning units to refrigerators to dishwashers. Residential motors are generally fractional horsepower motors, ranging in size from a few watts up to about 1 hp. Again, these are very rough classifications that will be used to analyze current motors and the possible benefits of more efficient motors.

The industrial motor load is the largest single category of electrical use in the country. At 679 billion kWh annually (1994 figure), industrial electric motors consume about 23% of all electricity generated in the United States. A very small overall increase in the efficiency of the industrial motor load would lead to enormous savings. For example, if the overall efficiency of industrial motor drives in the United States were increased by merely 0.1%, $30 million in annual electricity losses would be eliminated. Of course, larger increases in efficiency would translate into much larger savings.

The Office of Industrial Technology of the Department of Energy estimates that industrial motor use could be reduced by as much as 11 to 18% if all cost-effective efficiency technologies and practices were utilized; that translates to annual energy savings of 75 to 122 billion kWh. Note that these efficiency technologies and practices refer to motor efficiency upgrades (using a more efficient motor) as well as system efficiency measures (reducing the mechanical load, using speed controls, better maintenance, etc.). The same study estimates that the initial cost to implement these efficiency technologies and practices would be $11 to $17 billion in the form of capital expenditures. The annual savings would amount to between $3.8 and $5.8 billion. Thus, in three to four years, the improvements would pay for themselves.

Residential motor load is also a very large portion of the energy used in the United States. Approximately 445 billion kWh are consumed annually by the small motors used in residential applications. This is roughly one-quarter of all energy used by the residential sector. Like the industrial sector, the residential sector could also greatly benefit from more efficient motor drives. However, it is even more feasible to introduce more efficient motor drives to residential applications because the life spans of household appliances and gadgets are generally shorter than the life spans of large industrial machines. Thus there are more opportunities to replace inefficient motors. In addition, the price difference between highly efficient and typical motor drives is smaller at the residential level than it is at the industrial level. While a high-efficiency refrigerator may cost $20 to $50 more than a similar unit with lower efficiency, a large industrial high-efficiency motor drive may cost hundreds or even thousands more than less efficient motor drives. Therefore there are more opportunities to increase the efficiency of motor drives used in residential applications.

Power consumption and loss translates directly into fuel used in power generation. The slight increases in efficiency that could save customers millions of dollars a year in electricity could also conserve dwindling fossil fuel supplies and keep unnecessary pollutants out of the atmosphere. In the year 2000, 23.4 billion tons of CO_2 were released into the atmosphere as a result of worldwide power production. Since the electricity used by motors and motor drives accounts for approximately half of the world's power demand, it is logical to conclude that approximately 11 billion tons of CO_2 result from the use of all motors annually. To put this number in context, the entire Amazon rainforest absorbs about 2 billion tons of CO_2 annually. If all motor drives worldwide experienced an efficiency increase of 0.1%, approximately 10 million tons of CO_2 would not be released into the atmosphere. This is the amount of CO_2 absorbed by approximately 1 million hectares of temperate forest (which is one-tenth of all the forest land in Europe). When considering the direct but often overlooked correlation between energy use and pollution, the case for more efficient motor drives becomes much stronger. Table 6.1 lists potential global effects of motor drive efficiency improvements.

The number of electric motors, and hence the number of motor drives, is increasing. Fortunately, technological advances in power

TABLE 6.1 Global Effects of Motor Drive Efficiency Improvements

Global increase in motor drive efficiency	Approximate cost of electricity saved (based on average 5¢ kWh)	Approximate amount of CO_2 kept out of atmosphere
0.1%	$600 million	10 million metric tons
0.5%	$3 billion	50 million metric tons
1.0%	$6 billion	100 million metric tons

electronics, microprocessors, and electric motors are resulting in advanced motor drives that boost performance while increasing efficiency. The motor drive market is expected to grow from $12.5 billion in 2000 to $19 billion by 2005. This large market growth is responsible for making advanced motor drives more profitable and thus more common. The widespread use of high-efficiency, advanced motor drives creates the possibilities of reducing energy costs for end users, reducing the consumption of fossil fuels for energy production, and decreasing the amount of harmful emissions released into the atmosphere. The world has much to gain by researching advanced motor drives.

6.4 CURRENT MOTOR TECHNOLOGY

In today's modern world, electronics are everywhere, from handheld computers to air conditioners to projection TVs. However, even now over half of all power consumption in the United States can be accounted for by motors. These motors can vary from a simple blender and fan motor to an industrial motor used for assembly lines in automobile factories. When you consider the mass power that the United States consumes in a year, it becomes apparent that if one can make these motors run even a couple tenths of a percent more efficiently, it can make a huge difference in power savings. The purpose of this section is to first explain current motor technology and where efficiency efforts stand to date. The second purpose is to document future design innovations and their effect on overall efficiency.

Electric motors convert electrical energy into useful mechanical energy. This energy can then be used to drive household appliances, e.g., fans, compressors, etc.; but even in home applications, not all motors are alike. Different types have varying characteristics (and thus different efficiencies), making them suitable for certain situations but not for others. Single-speed induction motors are presently being used for most residential applications ranging from portable fans to compressors commonly found in refrigerators. These include both single- and three-phase squirrel-cage and shaded-pole induction motors. There is a noteworthy dissimilarity and a rather wide range of efficiency between these single-speed induction motors. Shaded-pole motors tend to be lower on the efficiency scale. It is also worth mentioning that in general motor efficiency comes at the price of horsepower, and for this reason smaller motors are generally less efficient.

Universal AC/DC motors are commonly used for sporadic applications where high speed is needed. Examples of this would be drills, food processors, and vacuums. These are the "brush motors" (named for the set of brushes that constantly change voltage, thus keeping the motor spinning in an attempt to align polarity). The main problem with these types is significant losses associated with the windings' wear and tear on the brushes (most small motors fail due to worn brushes).

Induction motors are found in refrigerators and air conditioners, but in many cases also in washing machines. It is in this area that a greater efficiency would yield huge returns in the long run.

6.5 ADVANTAGES OF VARIABLE-SPEED MOTORS

Most motors are designed to operate at a constant speed and provide a constant output. While in many cases this may be more than adequate, it is not in all. Two-speed induction motors can improve efficiency for refrigerators, air conditioners, and blowers. Although in theory this can be done with any induction motor application, a greater value is obtained with appliances that run frequently. With a two-speed mode of operation, long time periods that would normally use full power can be replaced by long periods of substantially less power with short periods when full power may be needed. Currently, residential central air conditioners, blowers (furnaces), and clothes washers take advantage of this technology since small changes in speed can drastically cut down on power usage (power consumption is approximately proportional to the cube root of shaft speed, e.g., a shaft reduction of 10% corresponds to at 27% reduction of power).

There are many ways to control the shaft speed of a motor. The most common way is via throttling devices such as valves and inlet vanes. However, this type of control is comparable to driving a car at a high speed and controlling the speed by using the brake. Another way is by using ASDs. This type of drive controls the speed by regulating the voltage, current, and/or frequency sent to the motor until the approximate load speed is obtained. Several types of ASDs are available, each with its own characteristics and practical applications. Even in these devices, there are many different kinds. Pulse width modulation (PWM) ASDs work by chopping pulses of varying widths to create the desired output voltage. They do this by using computer software which in turn is controlled by complex algorithms monitoring timing, duration, and frequency. This type of ASD has a rather high power factor, good response time, as well as low harmonic distortion. They also have the capability to contol many different motors from the same system. Their downfall is higher heat dissipation and a limited data cable length from the control to the motor.

Voltage source inverter (VSI) ASDs can also control many motors from a single drive and have the advantage of simple circuitry (an advantage that does not exist in PWM ASDs). They normally have a capacitor before the inverter to help store energy and keep the voltage stable. Their control ranges from about 10 to 200% of rated motor speed; however, below 10% it breaks down and becomes very inefficient.

The last common type of ASD is the current source inverter (CSI). It uses the inductive characteristics of the motor to stabilize DC as it reaches the inverter. Because this induction has to be rather large, this type of drive can only be used in medium to large motors. Advantages include short-circuit protection, quiet operation, and high efficiency at a wide range of speeds (normally above 50%). However, disadvantages include the inability to test the drive while not connected to a motor and complexity in connecting multiple motors to a single drive. Table 6.2 summarizes the characteristics of these four types of drives. Although many of these applications are beyond utilization for the small-scale motors being analyzed, it is mentioned to point out the potential energy saving for similar systems on an achievable smaller scale.

6.6 GOVERNMENT REGULATION

Recently, the federal government has made great attempts to strive for higher efficiencies in consumer goods. These attempts include

TABLE 6.2 Characteristics of Different Drives

	Pulse width modulation	Voltage source inverter	Current source inverter
Ease in retrofitting	Yes	Yes	No
Soft start	Yes	Yes	Yes
Regeneration	Option[a]	Option[a]	Inherent
Motor heating	High	Low	Low
Motor noise	High	Low	Low
Partial loading	Yes	Yes	Unstable
Low-speed operation	Smooth	No	No
Low-speed torque pulsation	No	Yes	Yes
Frequency above 60 Hz	Yes	Yes	No
Open-circuit protection	Inherent	Inherent	Required[b]
Short-circuit protection	Required[b]	Required[b]	Inherent
Overload protection	Required[b]	Required[b]	Inherent
Multimotor drive	Yes	Yes	Option[a]
Controller and logic	Complex	Simple	Semicomplex
60-Hz power factor	High	Poor	Poor
60-Hz harmonics	Low	High	High
Motor harmonics	High	Moderate	Moderate
Voltage stresses on motor	Yes	No	Yes
DC filter size	Small	Large	Large
Inverter noise	High	Medium	Medium
Transistor/GTO technology	Yes	Yes	No
Inverter switches	High frequency	Low frequency	Low frequency
Size and weight	Small	Medium	Small

[a] Feature is available at extra cost.
[b] Feature must be provided by the system design.
Source: Courtesy Pacific Gas and Electric Co.

minimum efficiencies that appliances must achieve. This has been done by to the passing of the National Appliance Energy Conservation Act (NAECA) in 1990. Since the passing of this act, appliance manufacturers have made great improvements, especially in refrigerators and freezers. It is for this reason that replacing an older refrigerator or freezer can lower the household energy usage by a significant amount. It has also been made law that all new major appliances show their efficiencies where the consumer can easily see them when making their purchase decision. This is done to try to alert the consumer of the importance of energy savings and its effect on goods. It also serves to show the consumer that they cannot ignore the fact that the energy cost of running an appliance can certainly outweigh its cost.

We now review the few cases where motor improvements create huge savings in the long run. A case for appliance replacement is furnace fans. Although newer models are much more efficient, most residents are using older models. These models can use quite a lot of energy. Right now most forced air systems and air conditioners use multiple-speed, shaded-pole, or permanent–split capacitor induction motors. The efficiency of these is in the 50–60% range when a single speed is selected to match the rest of the system. In these cases, there are not one or two but three options that may be used to improve efficiency. They include a high-efficiency motor, an ECM (electronically commutated permanent magnet) motor, and a variable-speed ECM motor. These can add anywhere from $15 to $75 to the retail cost of the system; however, they quickly pay for themselves in energy savings and will continue to lower energy costs. The case is very similar for heat pump blowers as well.

Air conditioners have also been through many changes since efficiency regulation. Their levels have risen from 10.5 to 11.5 only in the last decade. However, traditional motor improvement is coming to 2–3% of its practical limit (in the case of the compressor motor). This leaves very little room for traditional improvements; for example, if a compressor with 80% efficiency is being driven by a 90% efficient motor, the new rating would by only 11.8, a small difference considering the work involved in achieving the compressor and motor efficiencies. This brick wall is mainly due to the fact that air conditioners primarily operate at a single speed, cycling on and off to meet the current load required (some also control output with valves and dampers; at times these may be even less efficient). The solution is a variable-speed motor that would operate at highly variable loads by matching the speed to the load. This would save energy and extend the life of the motor by allowing it to operate less often at full throttle. Because applying this strategy would in-

TABLE 6.3 Effects of Efficient Motor Options for Indoor Blowers

Motor	Efficient motor option	Energy savings, %	Energy savings, $/yr	Additional retail cost, $	Simple payback, yr
Central A/C blower	High-efficiency PSC	14	4.50	15	3
	ECM	25	8.00	40	5
	Variable-speed ECM	75	24.00	75[a]	3
Heat pump blower	High-efficiency PSC	14	11.20	15	1
	ECM	25	20.00	40	2
	Variable-speed ECM	75	60.00	75[a]	1
Furnace blower	High-efficiency PSC	14	6.70	15	2
	ECM	25	12.00	40	3
	Variable-speed ECM	75	36.00	175[b]	5
Central A/C and furnace blower	High-efficiency PSC	14	11.20	15	1
	ECM	25	20.00	40	2
	Variable-speed ECM	75	60.00	175[b]	2

[a] Cost of variable-speed blower only.
[b] Includes incremental cost of $100 for capacity modulation in the furnace.
Source: homeenergy.org.

volve more than just improving efficiency in an already designed air conditioner—in fact, in most cases it involves redesigning the entire unit as a whole—manufacturers are very reluctant to take this approach.

Residential refrigerators and freezers can have up to three motors (excluding ice makers and defrost timers); the largest drives the refrigerant compressor. In frost-free units (the most common), two additional motor units drive fans that circulate air over the condenser and evaporator. The compressor is normally an AC single-phase, two-pole induction motor. In order to comply with a dramatically reduced allowable refrigerator and freezer energy con-

TABLE 6.4 Potential for Residential Energy Savings Through Increased Motor Efficiency

Application	Total national motor energy use, 10^9 kWh/yr	Current motor efficiency savings, %	Practical efficiency,[a,b] %	Potential energy %	Potential energy 10^9 kWh	Typical payback,[b] yr
Refrigerators and freezers						
Compressor	101	80	82–84	4	4.0	14
Condenser fan	6	15	65	77	4.6	6
Evaporator fan[c]	6	15	65	128	7.7	4
Central A/C and heat						
Pump compressor	159	87	90	3	5.5	16
Outside unit fan	21	50	70	29	6.1	6
Room A/C compressor	25	87	90	3	0.8	13
Indoor A/C and heating blowers	61	60	80	25	15.3	3
Clothes washer motor	10	65	75	13	1.3	10

[a] Based on upgrading installed motor base to maximum practical efficiency levels.
[b] Assuming average electric rate of $0.08/kWh.
[c] Includes reduction in compressor load.
Source: Opportunities for Energy Savings in the Residential and Commercial Sectors with High Efficiency Electric Motors, U.S. Department of Energy.

sumption by the federal government in 2001 and even more reductions foreseeable in the near future, manufacturers are now considering ECMs. These would adjust to load conditions either at a constant speed or at a variable speed. Either of these options would be a great improvement over current practices in which the compressor motor runs at full speed until the desired temperature is met, then cycles on and off between full speed and off.

It is estimated that replacing both the condenser motor and fan motor would increase the retail price by only $75; but the consumer would see the improvement pay for itself in a few years. Table 6.3 shows the efficient motor options for indoor blowers. Table 6.4 shows the potential for savings by increased motor efficiency. And Table 6.5 displays the potential for saving by variable speed motors.

TABLE 6.5 Potential for Residential Energy Savings Through Variable-Speed Motors

Application	Total national motor energy use, 10^9 kWh/yr	Current motor efficiency, %	Practical efficiency,[a] %	Energy savings		Typical payback,[b] yr
				%	10^9 kWh	
Refrigerator/freezer compressor	101	80	88	20	20.2	8
Central A/C and heat pump compressor[c]	159	87	90	35	55.7	15–25
Room A/C compressor	25	87	90	10	2.5	20+
Indoor heating and A/C blowers	61	60	80	75	45.9	2–3

[a] Based on upgrading installed motor base to maximum practical efficiency levels.
[b] Assuming average electric rate of $0.08/kWh.
[c] Using a two-speed induction motor. (Somewhat higher energy savings are possible with a continuously variable-speed motor, but the payback period is longer.)
Source: Opportunities for Energy Savings in the Residential and Commercial Sectors with High Efficiency Electric Motors. U.S. Department of Energy.

6.7 ADJUSTABLE-SPEED DRIVE APPLICATIONS

Adjustable-speed drives can be used in every environment where variable speed and torque are needed. ASDs are currently used in elevators, water and wastewater pumps, boiler fans, HVAC systems, wind turbines where speed of wind is not constant, hydroelectric plants where speed of water is not constant, and the automobile industry. Some of these applications will be discussed in detail below.

Over the last few years there has been an increase in demand for electrical energy that has been made by wind energy, an alternative energy source. Wind speed is variable and applies variable speeds to the generator at the turbine, which creates different frequency levels. To increase productivity and get 60-Hz frequency at the output, adjustable-speed drives can be used in wind turbines. In advanced systems, wind is converted to electrical energy via an AC generator with variable frequency and voltage levels. By using a converter, this AC is converted to DC; then, using a DC link line, this DC is transmitted to a DC/AC inverter. The AC utility line has fixed voltage and frequency with better power quality. Illinois, Texas, and Arizona in the United States as well as Denmark and others use wind energy. Other countries are sure to follow.

As with wind energy, water flow also has variable speed. This again creates variable voltage, frequency, and a poor power factor. By using ASDs, these problems can be fixed and the AC utility line can have fixed values. In such advanced systems, water flow is converted into electrical energy via an AC generator with variable frequency and voltage levels. By using a converter, this AC is converted to DC; then, using a DC link line, this DC is transmitted to an inverter connected to the AC utility line, which has fixed voltage and frequency with better power quality. This increases quality and efficiency. Hydroelectric power plants are one of the oldest and biggest electrical energy creation systems.

Centrifugal pump and fan applications are among of the most common areas for ASDs. Gas or liquid flow can be regulated using ASDs. Adjustable-speed drives can reduce maintenance costs and increase the efficiency of pumps to almost the best efficiency point.

In Table 6.6, the relationships among the speed of a pump, liquid flow, and horsepower needs are shown. As seen in the table, increments in the flow and speed will greatly decrease power requirements. If the speed of the pump decreases by 50%, the power requirements will go down to 13%. This shows that ASDs can be very efficient and save money.

In conclusion, ASDs and their applications have been rapidly growing over the last few years. There are many applications, like fans, compressors, pumps, automobile industry applications, and many other motor areas, where variable-speed drives are desirable and can be readily found. There are some downsides to ASDs, such as harmonics, voltage notching, and complexity of design and cost, but these can be fixed using harmonic filters, voltage regulators, and switching control mechanisms. Use of ASDs will increase the efficiency of machinery, lower maintenance costs, and reduce machine size. Faster-switching semiconductor devices and a decrease in semiconductor prices make this field very popular for engineers and the industry. Available ASDs can be purchased as AC or DC drives. AC drives can adjust the frequency and voltage to change the speed of the electric machine. Creating a voltage and frequency ratio constant will create a constant torque. These drives are newer and simpler compared to the DC drives, but they are not as efficient. AC machines need less maintenance than DC machines and motor speed can reach almost four times more than DC machines. Another advantage of using vector-controlled AC machines is that

TABLE 6.6 Flow–Speed–Power Relationships of Typical Pumps

Flow, %	Speed, %	Horsepower, %
30	30	3
50	50	13
70	70	34
80	80	51
90	90	73
100	100	100

fast-changing load applications can be used. DC drives need converters to convert AC utility voltage into DC voltage, and by changing this the voltage speed of the machine can be altered. DC machines require better maintenance and therefore its environment needs to be clean and dry. Adjustable-speed drives are one of the newest and fastest growing technologies in electric machines, and their future will certainly be dynamic.

SELECTED READINGS

1. Krishnan, R. (2001). *Electric Motor Drives: Modeling, Analysis, and Control.* Upper Saddle River, NJ: Prentice-Hall.
2. El-Sharkawi, A. (2000) *Fundamentals of Electric Drives.* Pacific Grove, PA: Brooks/Cole Publishing.
3. Mohan, N., Undeland, T. M., Robbins, W. P. (2003). *Power Electronics: Converters, Applications, and Design.* New York: John Wiley & Sons.
4. Bose, B. K. (2002). *Modern Power Electronics and AC Drives.* Prentice Hall PTR.
5. Mohan, N. (2001). *Electric Drives: An Integrative Approach.* Minneapolis: MNPERE.
6. Mohan, N. (2001). *Advanced Electric Drives.* Minneapolis: MNPERE.
7. Skvarenina, T. L. (2002). *The Power Electronics Handbook.* Boca Raton, FL: CRC Press.
8. Taking Control of Energy Use. *Home Energy Magazine Online.* May/June 1998.
9. United States Industrial Electric Motor Systems Market Opportunities Assessment, Office of Industrial Technologies (OIT), Dec. 2002.
10. Motors Matter. *Home Energy Magazine Online*, July/Aug. 2000.
11. Key World Energy Statistics, International Energy Agency (IEA), 2003.
12. Accelerated Global Warming and CO_2 Emissions. *Hydrogen Now! Journal*, Issue 2, Article 1a, 2003.
13. Electronic Motor Drives 2001–2005, Report #1202, Drives Research Corporation, Aug. 2001.

7

Induction Motors and Adjustable-Speed Drive Systems

7.1 ENERGY CONSERVATION

The potential energy conservation in any system can best be determined by examining each element of the system and its contribution to the losses and inefficiency of the system. Every device that does any work or causes a change in the state of a material has energy losses. Thus, typical losses include the following:

1. Electrical transmission losses from the metering point to the system. (This is where the electric power consumption is measured and the power bill determined.)
2. Conversion losses in any power conditioning equipment. (This includes variable-frequency inverters and the effect of the inverter output on the motor efficiency.)
3. Electric motor losses to convert electric power to mechanical power.

4. Mechanical losses in devices such as gears, belts, and clutches to change the output speed of the motor.
5. Losses in the driven unit, such as a pump or fan or any other device that performs work on material.
6. Transmission losses, such as friction losses to move material from one location to another.
7. Losses caused by throttling or other means to control material flow by absorbing or bypassing excess output.

Each element in a particular system has an efficiency that can be defined as

$$E = \frac{\text{output power}}{\text{input power}}$$

or

$$\text{Losses} = \text{input power} - \text{output power}$$

The overall efficiency of the system is the product of the efficiencies of all elements of the system; thus,

$$E_{overall} = E_1 \times E_2 \times E_3 \times E_4 \times E_5$$

Therefore, the proper selection of each element can contribute to electric energy conservation.

This can be illustrated by an example of a constant-speed pumping system. The pumping system is to move water from one location to another at 1000 gpm with a static head of 100 ft. The friction head is 30 ft with a 4-in.-diameter supply pipe. What is the energy saving using a 5-in.-diameter supply pipe and an energy-efficient motor? Table 7.1 shows a summary of the calculations for the two systems. The net result is an annual saving of 28,520 kWh, or 19.7% of the input. Note that the savings were achieved by improved performance in several elements: lower motor losses due to improved efficiency and lower horsepower required, lower pump losses due to increased efficiency and lower horsepower required,

TABLE 7.1 Summary Calculations for Example of Pump Installation

	4-in. supply pipe	5-in. supply pipe
Static head, ft	100	100
Friction head, ft	30	10
Total head, ft	130	110
Output hydraulic hp	25.25	25.25
Input hydraulic hp	37.83	27.78
Hydraulic efficiency	0.769	0.909
Pump efficiency	0.77	0.79
Pump input, hp	42.64	35.16
Motor standard hp	50	40
Motor efficiency at operating load	0.90	0.92
Motor input, hp	47.38	38.22
Transmission efficiency	0.978	0.982
System efficiency	0.521	0.649
System input, hp	48.46	38.91
System input, kW	36.15	29.02
Energy saving, kW	—	7.13
Annual saving, kWh	—	28,520
Percent savings	—	19.7

and lower pipe friction losses with improved hydraulic efficiency. The overall system efficiencies are as follows:

$$4\text{-in. system efficiency} = 0.769 \times 0.77 \times 0.90 \times 0.978$$
$$= 0.521$$

$$5\text{-in. system efficiency} = 0.909 \times 0.79 \times 0.92 \times 0.982$$
$$= 0.649$$

The conclusion is that the complete system needs to be considered to obtain the most energy-efficient installation. One aspect not to be overlooked is that the losses in the system are dissipated as heat at each device, such as the motors, pumps, and compressors. Therefore, if the devices are in a conditioned environment, the effect of the losses or change in the losses on the conditioning system must also be considered.

In many installations, additional energy savings can be achieved by combining the fixed-speed induction motor with some method of varying the output speed of the unit. This is particularly true in any application in which output is a fluid flow that must vary in response to some other variable. A similar opportunity exists if output pressure must be controlled with a varying flow or varying input pressure.

Many fluid processes (including air processes) involve pumping the fluid to a high pressure and controlling flow and pressure to the required levels by throttling or bypassing. These throttling and bypass methods of control are inherently inefficient.

Centrifugal pumps, fans, and blowers have characteristics in accordance with the laws of fan performance, which state the following:

Flow varies directly with speed.
Pressure varies as the square of the speed.
Power varies as the cube of the speed.

These types of applications lend themselves to conversion from throttled constant-speed systems to adjustable-speed systems and offer a large potential for energy savings.

7.2 ADJUSTABLE-SPEED SYSTEMS

Many types of adjustable-speed systems are available. Some of the more popular types of adjustable-speed drives are the following: multispeed motors, adjustable-speed pulley systems, mechanical adjustable-speed systems, eddy current adjustable-speed drives, fluid drives, DC adjustable-speed systems, AC variable-frequency systems, and wound-rotor motors.

The selection of the most effective system for a specific application depends on a number of factors:

Life-cycle cost
First cost

Duty cycle and horsepower range
Energy consumption
Control features required
Size
Performance
Reliability
Maintenance

To assist in the selection of an adjustable-speed drive system, let us examine the characteristics of the more popular ones. The DC adjustable-speed systems have been specifically excluded from this section since their characteristics and application technology are well known to those who apply them.

7.2.1 Multispeed Motors

As discussed earlier, multispeed motors can be obtained with the following output characteristics:

Constant horsepower
Constant torque
Variable torque

However, in conventional multispeed motors, only a limited variety of speed combinations is available. One-winding, two-speed motors are available with 2-to-1 speed combinations such as

1750 rpm/850 rpm
1150 rpm/575 rpm

Two-winding, two-speed motors are available with speed combinations other than 2 to 1. Typical speed combinations are

1750 rpm/1150 rpm
1750 rpm/850 rpm
1750 rpm/575 rpm
1150 rpm/850 rpm

Thus, there are more combinations of speed ratios available in the two-winding, two-speed motors.

In addition, two-winding, four-speed motors are also available. A typical speed combinations is 1750 rpm/1150 rpm/850 rpm/575 rpm.

Since the power requirements for many fans and centrifugal pumps are a cube of the speed, the variable-torque multispeed motor can be used for two-step speed control, i.e., a high-speed and a low-speed operation. With a one-winding, two-speed motor, the output of the fan or pump on low speed will be 50% of the output on high speed, and the horsepower required will be 12.5% of high speed. Figure 1.7 shows a fan load curve superimposed on the speed-torque curves for a variable-torque multispeed motor. In the case of a two-winding, two-speed motor with a combination of 1750 rpm/1150 rpm, the output of the fan or pump on low speed will be 67% of the output on high speed, and the horsepower required will be 30% of high speed. This is illustrated by Fig. 1.8, which shows a fan load curve superimposed on the motor speed-torque curves. Given the speed limitations of this type of drive, it is an economical and reliable method to obtain incremental flow control.

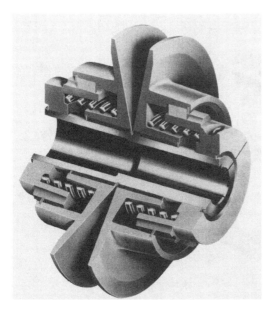

FIGURE 7.1 Variable-speed pulley. (Courtesy T. B. Wood's Sons Company, Chambersburg, PA.)

7.2.2 Adjustable-Speed Pulley Systems

An adjustable-speed pulley system consists of the electric motor mounted on a special base, an adjustable-speed sheave on the motor shaft, and a fixed-diameter sheave on the load shaft connected by a V belt. Figure 7.1 shows the construction of one of the variable-speed sheaves, and Fig. 7.2 shows the special base required for the drive motor.

Figure 7.3 illustrates the cross-section construction of a double-side spring sheave, showing the maximum and minimum drive-belt locations, and Fig. 7.4 shows the cross-section construction of a single-side spring sheave.

The double-side spring sheave is usually recommended for the integral-horsepower V-belt drives. The V belt in the spring-loaded sheave changes its diametric position as the base is adjusted, resulting in a change in the ratio of the effective pitch diameters of the driven and driving sheaves and a change in output speed. This type of drive has a limited capacity up to approximately 125 hp and

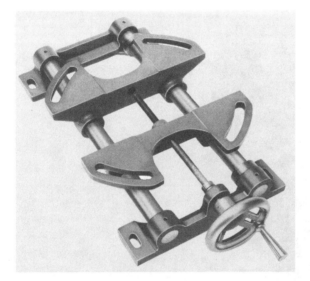

FIGURE 7.2 MBA motor base. (Courtesy T. B. Wood's Sons Company, Chambersburg, PA.)

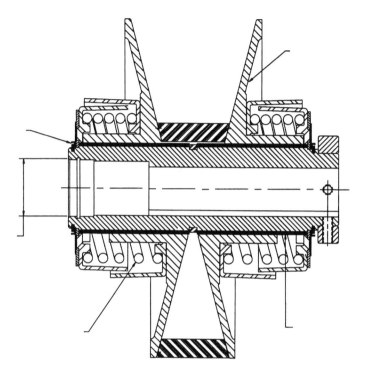

FIGURE 7.3 Cross section of double-side spring sheave. (Courtesy Lovejoy Inc., South Haven, MI.)

a limited speed range of approximately 2 or 3 to 1. In special cases, the speed range may be wider. The efficiency of these drives depends on the belt loading and minimum diameter of the sheaves. The belt efficiency as a function of load ranges from 92 to 99% and as a function of sheave diameter from 91 to 97%. When combined with a three-phase induction motor, the system efficiency ranges from 40 to 90%, depending on the type of load and speed reduction.

Figure 7.5 illustrates the system efficiency for an adjustable-speed pulley system with a 10-hp, 1750-rpm energy-efficient motor driving a constant-torque load over a 3-to-1 speed range.

Figure 7.6 illustrates the system efficiency for an adjustable-speed pulley system with a 10-hp, 1750-rpm energy-efficient motor

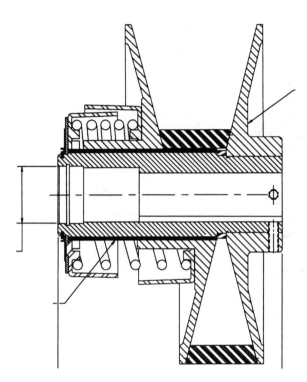

FIGURE 7.4 Cross section of single-side spring sheave. (Courtesy Lovejoy Inc., South Haven, MI.)

driving a variable-torque load ovr a 3-to-1 speed range. Its major disadvantage is that the speed must be changed manually and does not lend itself to automatic or remote control. However, for applications that require only occasional adjustment in output, this system may be adequate and still provide energy savings at the lower speed settings, for example, an air-handling system that requires output adjustment only for summer and winter operation. Figure 7.7 shows a typical installation of this type of drive.

7.2.3 Mechanical Adjustable-Speed Systems

The broad group of mechanical ajustable-speed drives includes the more common stepless mechanical adjustable-speed drives that

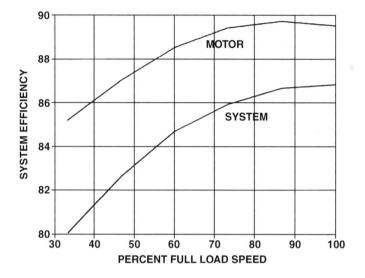

FIGURE 7.5 Adjustable-speed pulley system with a 10-hp energy-efficient motor driving a constant-torque load.

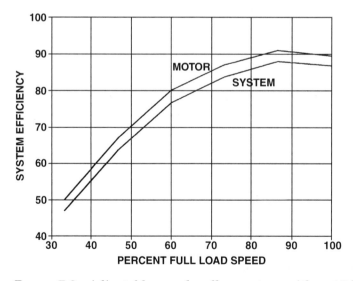

FIGURE 7.6 Adjustable-speed pulley system with a 10-hp energy-efficient motor driving a variable-torque load.

FIGURE 7.7 Installation of variable-speed sheave. (Courtesy T. B. Wood's Sons Company, Chambersburg, PA.)

provide an infinite number of speed ratios within a nominal speed range. These type of drives include packaged belt and chain drives, friction drives, and traction drives. These drive systems are usually driven by a constant-speed induction motor and convert this constant-speed input into a stepless variable-speed output.

Figure 7.8 is a cross section of the assembly of the U.S. Electric Motors varidrive system, showing the electric motor, driver, driven sheaves, and speed-adjusting mechanism.

A typical group of packaged adjustable-speed belt-drive systems is shown in Fig. 7.9. In the case of the belt-drive systems, the basis of rating is generally constant torque (variable horsepower) at speed ratios below 1 to 1 and constant horsepower (variable torque) at speed ratios above 1 to 1. Figure 7.10 illustrates this basis of

FIGURE 7.8 Cross section of U.S. Electrical Motors Varispeed System. (Courtesy U.S. Electrical Motors, Division of Emerson Electric Co., St. Louis, MO.)

FIGURE 7.9 U.S. Electrical Motors varidrive units. (Courtesy U.S. Electrical Motors, division of Emerson Electric Co., St. Louis, MO.)

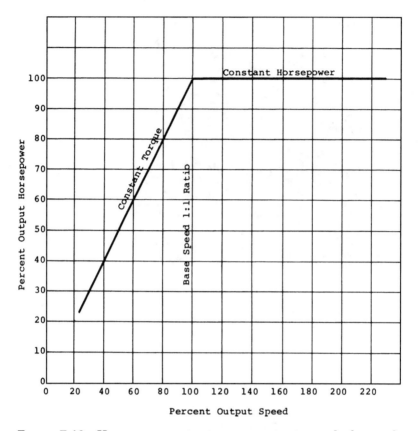

FIGURE 7.10 Horsepower output versus output speed of a mechanical adjustable-speed drive system. (Courtesy U.S. Electrical Motors, division of Emerson Electric Co., St. Louis, MO.)

rating. The efficiency of these systems at various loads and speeds is shown in Figs. 7.11 and 7.12 for representative ratings of the varidrive line of packaged mechanical belt drives.

Most of these types of drives have limited horsepower and speed ranges. Therefore, the selection of the drive systems should be based on the duty cycle of the load and the characteristics of the drive under consideration, including speed range, horsepower, torque characteristics, and efficiency over the duty cycle. When

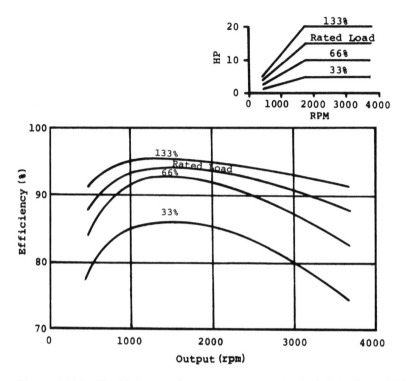

FIGURE 7.11 Varidrive performance curves, typical data for a 15-hp, four-pole motor. (Courtesy U.S. Electrical Motors, division of Emerson Electric Co., St. Louis, MO.)

properly applied, many of these drives have good efficiencies over their operating range. It is recommended that several types of systems be compared to determine the most suitable and effective life-cycle cost system. The requirements and type of remote control must also be a factor in the selection of the drive system.

Figures 7.13–7.15 illustrate the types of process controls available on packaged belt drives.

7.2.4 Eddy Current Adjustable-Speed Drives

The operating principle of the eddy current drive system involves a constant-speed AC induction motor that is magnetically coupled to

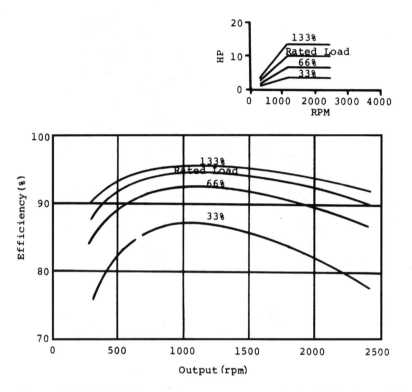

FIGURE 7.12 Varidrive performance curves, typical data for a 10-hp, six-pole motor. (Courtesy U.S. Electrical Motors, division of Emerson Electric Co., St. Louis, MO.)

an output shaft through an integral variable-speed eddy current coupling. The eddy current coupling consists of a constant-speed drum that is directly connected to the drive motor rotor and an inductor that is directly connected to the output shaft. As the drum rotates, eddy currents are induced and magnetic attraction occurs between the drum and the inductor, thus transmitting torque from the constant-speed drum to the output inductor. An excitation winding, which is usually stationary, is located in the magnetic circuit and is excited by a DC current to provide the magnetic field in the constant-speed drum and variable-speed inductor. The application of the field current creates a magnetic flux across the air gap

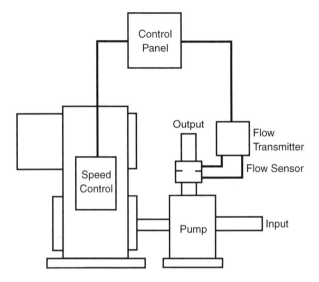

FIGURE 7.13 Output flow control of a mechanical adjustable-speed system.

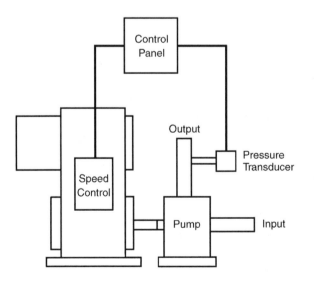

FIGURE 7.14 Line pressure control of a mechanical adjustable-speed system.

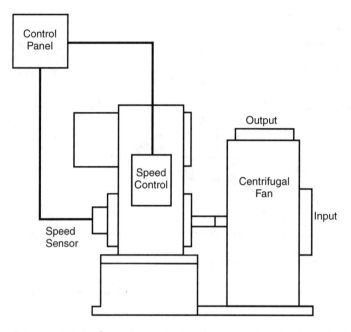

FIGURE 7.15 Speed control of a mechanical adjustable-speed system.

between the two members of the clutch, which induces eddy currents in the input drum. The net result is a torque available at the output shaft. The variation in the field current varies the degree of magnetic coupling between the motor-driven constant-speed drum and the variable-speed inductor connected to the output drive shaft. By adjustments in the field current, the output speed can be adjusted to match the output load requirements (speed and torque). Figure 7.16 shows the cross-section assembly of a self-contained eddy current drive system. Figure 7.17 is a general view of such a system. The power flow for this type of adjustable-speed drive is shown in Fig. 7.18.

The degree of coupling or slip between the two members is determined by the load and level of excitation. The slipping action (i.e., difference in speed) is the source of the major power loss and

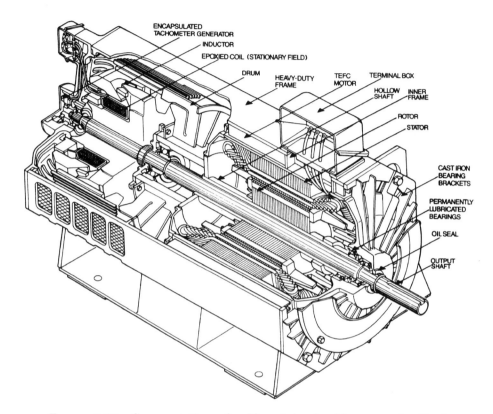

FIGURE 7.16 Cross section of self-contained eddy current adjustable-speed drive system. (Courtesy Magnetek, New Berlin, WI.)

inefficiency of the eddy current coupling. This slip loss is the product of the slip rpm, which is the difference in speed between the input and output members and the transmitted torque. This relationship may be expressed as follows:

$$\text{Load (output) hp} = \frac{\text{rpm}_2 T_L}{5252}$$

$$\text{Motor (input) hp} = \frac{\text{rpm}_1 T_L}{5252}$$

FIGURE 7.17 Self-contained eddy current adjustable-speed drive system. (Courtesy Magnetek, New Berlin, WI.)

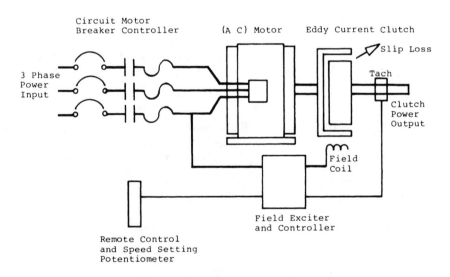

FIGURE 7.18 Power flow for eddy current drive system.

Slip loss = motor hp − load hp

$$= \frac{T_L(\text{rpm}_1 - \text{rpm}_2)}{5252}$$

where

rpm_1 = coupling input speed (motor)
rpm_2 = coupling output speed (load)
T_L = load torque, ft-lb

The efficiency of an eddy current coupling can never be greater than the numerical percentage of the output speed. However, in addition to the slip losses, the friction and windage losses and excitation losses of the coupling must also be included in the efficiency determination. The friction and windage loss is about 1% of the rated input horsepower and can be considered constant over the speed range. The excitation loss is less than 0.5% of the input horsepower and decreases with reduction in speed. The maximum torque developed by the drive is limited to the maximum torque (breakdown torque) of the induction drive motor or the magnetic coupling of the eddy current clutch. With proper matching of the drive components, the full capacity of the drive motor can be utilized. Figure 7.19 illustrates the overload capacity of an eddy current drive system with an induction motor driver.

Since the eddy current coupling has no inherent speed regulation, it is necessary that the coupling include a tachometer generator that rotates at the coupling output speed. The tachometer-generator output signal is fed into a speed-control loop in the excitation system to provide close output speed regulation. The speed regulation is usually ±1% but with closed-loop control may be as close as ±0.1%. In addition to speed, the eddy current-control system can be used with any type of actuating device or transducer that can provide a mechanical translation or an electrical signal. Actuating devices include liquid-level control, pressure control, temperature control, and flow control. The performance of the eddy current adjustable-speed drive system driving a constant-torque load is illustrated in Fig. 7.20. The speed range for continuous operation is usually 16:1 but can be wider, depending on the thermal dissipa-

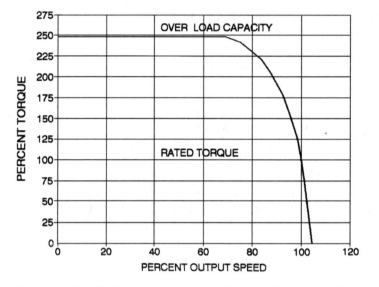

FIGURE 7.19   Eddy current adjustable-speed drive overload capacity.

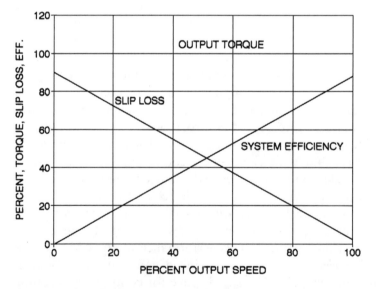

FIGURE 7.20   Eddy current adjustable-speed drive applied to a constant-torque load.

tion capacity of the eddy current clutch. The performance of the eddy current adjustable-speed drive system driving a variable-torque load is illustrated in Fig. 7.21. The speed range shown in Fig. 7.21 is down to only 45% speed, which is within the range of most variable-speed applications; however, the eddy current drive can supply these types of loads to much lower speeds if necessary.

The advantages of this type of adjustable-speed drive are

Eddy current simplicity and high reliability
Stepless variable-speed control
Good speed regulation
High starting torque
High overload capacity
Controlled acceleration
Handle high-impact loads.

Figure 7.22 illustrates the application of an eddy current adjustable-speed drive to an extruder. Figure 7.23 illustrates the application of the eddy current adjustable-speed drive to a process pump.

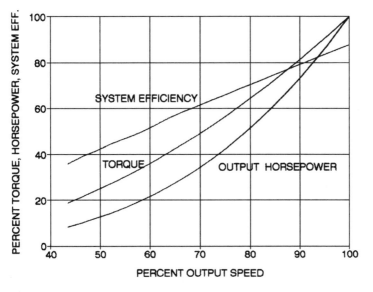

FIGURE 7.21 Eddy current adjustable-speed drive applied to a variable-torque load.

FIGURE 7.22 The application of an eddy current adjustable-speed drive system to an extruder. (Courtesy Magnetek, New Berlin, WI.)

7.2.5 Fluid Drives

Fluid drives can be described as any device utilizing a fluid to transmit power. The fluid generally used is a natural or synthetic oil. Fluid drives can be grouped into four categories: (1) hydrokinetic, (2) hydrodynamic, (3) hydroviscous, and (4) hydrostatic. The hydrokinetic, hydrodynamic, and hydroviscous drives are all slip-type devices.

The hydrokinetic fluid drive, commonly referred to as a fluid coupling, consists of a vaned impeller connected to the driver and a vaned runner connected to the load. The oil is accelerated in the impeller and then decelerated as it strikes the blades of the runner. Thus, there is no mechanical connection between the input and output shafts. Varying the amount of oil in the working circuit changes the speed. This provides infinite variable speed over the operating range of the drive. Figure 7.24 is a representation of such a drive.

FIGURE 7.23 The application of an eddy current adjustable-speed drive system to a process pump. (Courtesy Magnetek, New Berlin, WI.)

The circulating pump, driven from the input shaft, pumps oil from the reservoir into the housing through an external heat exchanger and then back to the working elements. The working oil, while it is in the rotating elements, is thrown outward, where it takes the form of a toroid in the impeller and runner. Varying the quantity of oil in this toroid varies the output speed. A movable scoop tube controls the amount of oil in the toroid. The position of the scoop tube can be controlled either manually or with automatic control devices. The scoop-tube adjustment gives a fast response and smooth stepless speed control over a wide speed range, i.e., 4 to 1 with a constant-torque load and 5 to 1 with a variable-torque load. In addition to providing speed control, the fluid drive limits torque and permits no-load starting on high-inertia loads.

These units range in size from 2 to 40,000 hp, as illustrated in Fig. 7.25.

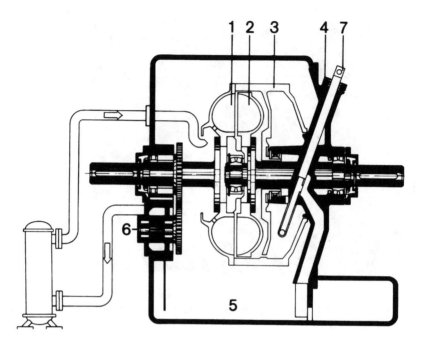

FIGURE 7.24   Diagram of a hydrokinetic drive: (1) primary wheel; (2) secondary wheel; (3) shell; (4) scoop tube housing; (5) oil sump; (6) oil pump; (7) scoop tube. (Courtesy Voith Transmissions, Inc., York, PA.)

Efficiency.   The fluid drives have two types of losses:

> *Circulation losses.* These losses include friction and windage losses, the power to accelerate the oil within the rotor, and the power to drive any oil pumps that are part of the system. These losses are relatively constant and are approximately 1.5% of the unit rating.
> *Slip losses.* As in the case of eddy current couplings, the torque at the input shaft is equal to the torque required at the output shaft:

$$\text{Load (output) hp} = \frac{\text{rpm}_2 T_L}{5252}$$

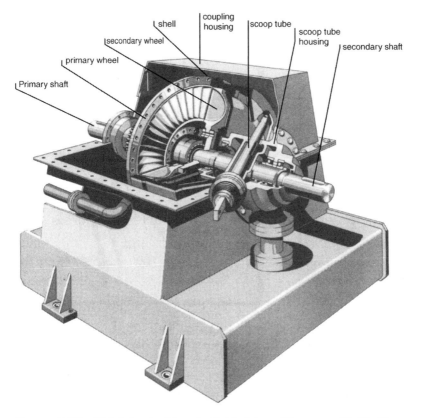

shell
coupling housing
scoop tube
scoop tube housing
secondary shaft
secondary wheel
primary wheel
Primary shaft

FIGURE 7.25 Voith hydrokinetic fluid drive. (Courtesy Voith Transmissions, Inc., York, PA.)

$$\text{Motor (input) hp} = \frac{\text{rpm}_1 T_L}{5252}$$

$$\text{Slip loss} = \text{motor hp} - \text{load hp}$$

$$= \frac{T_L(\text{rpm}_1 - \text{rpm}_2)}{5252}$$

where

$$\text{rpm}_1 = \text{coupling input speed (motor)}$$

rpm_2 = coupling output speed (load)

T_L = load torque, ft-lb

The slip efficiency is then

$$\frac{\text{Input hp} - \text{slip loss}}{\text{Input hp}} \times 100$$

$$= \frac{(T_L \text{rpm}_1/5252) - (T_L/5252)(\text{rpm}_1 - \text{rpm}_2)}{T_L \text{rpm}_1/5252} \times 100$$

$$= \frac{\text{rpm}_2}{\text{rpm}_1} \times 100$$

$$\text{Total input hp} = \frac{\text{output hp}}{\text{slip eff}} + \text{circulation hp losses}$$

$$\text{Coupling efficiency} = \frac{\text{output hp}}{\text{input hp}} \times 100$$

The maximum speed of the fluid drive at full load is about 98% of the driving motor speed, and with circulation losses of 1.5% the maximum efficiency is 96.5% at a maximum speed.

Figure 7.26 illustrates the typical performance of a fluid coupling driving a variable-torque load such as a fan or pump, where the torque varies as the speed squared and the horsepower varies as the speed cubed. Figure 7.27 illustrates the performance of a fluid coupling driving a constant-torque load such as a conveyor or piston pump, where the horsepower varies as the speed. Figure 7.28 illustrates a complete-package adjustable-speed fluid drive consisting of the drive motor, fluid coupling, and necessary accessories. Figure 7.29 shows the installation of a variable-speed fluid-drive system driving mud pumps at a mining installation. Figure 7.30 shows the installation of variable-speed fluid-drive systems driving blowers.

More complex units are available at ratings generally above 1000 hp. The Voith MSVD multistage variable-speed drives are an example of these drives, which consist of

Hydrodynamic variable-speed coupling
Hydraulic-controlled lock-up clutch

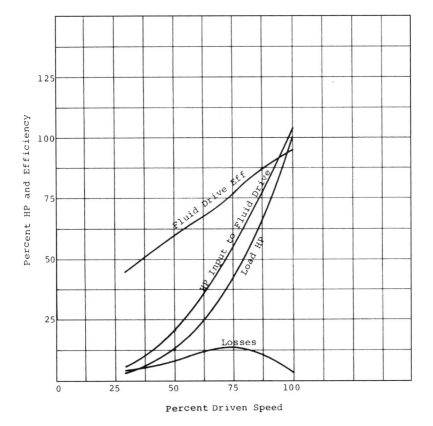

FIGURE 7.26 Fluid-coupling variable-speed drive characteristics when driving a load that varies as the speed cubed.

    Hydrodynamic torque converter
    Hydrodynamic brake
    Planetary gear, fixed
    Planetary gear, revolving

The operation of these units can be divided into two stages. In stage 1, the power is transmitted by the hydrodynamic variable-speed coupling directly through the planetary gear. The speed is controlled by changing the level of the oil in the hydrodynamic coupling. The

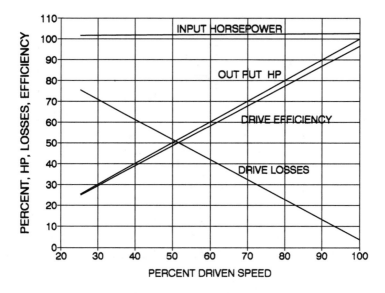

FIGURE 7.27 Fluid-coupling variable-speed drive characteristics when driving a constant-torque load.

FIGURE 7.28 Packaged fluid drive consisting of the drive motor, fluid coupling, and necessary accessories. (Courtesy Voith Transmissions, Inc., York, PA.)

FIGURE 7.29 Variable-speed fluid drives driving mud pumps at mining installation. (Courtesy Voith Transmissions, Inc., York, PA.)

FIGURE 7.30 Variable-speed fluid drives driving blowers. (Courtesy Voith Transmissions, Inc., York, PA.)

operating range is approximately 0–80% speed. The torque converter has no function in this stage. The hydrodynamic brake generates the countertorque for the planetary gear. In stage 2, the impeller and turbine wheel on the hydrodynamic coupling are locked together by the hydraulic-controlled clutch bridging the input and output elements so that the drive motor is now coupled mechanically to the driven load. The operating speed range in this stage is 80–100% and is controlled by the hydrodynamic torque converter.

Hydrostatic Drives. A hydrostatic variable-speed drive consists of a positive-displacement hydraulic pump driven by an induction motor, a positive-displacement hydraulic motor, and necessary hydraulic controls. The hydraulic pump and motor are usually separate units. This type of drive is also offered as a package consisting of the hydraulic pump, the piping, and the hydraulic motor mounted in a common housing.

When the hydraulic pump is driven by a constant-speed AC induction motor, the variable output is obtained by controlling the speed of the hydraulic motor. Commonly, the easiest system to design may be the most energy inefficient. Throttling any valve in the hydraulic system generates heat and consumes energy. The significance of this power loss is expressed as follows:

$$\text{Power loss (hp)} = \frac{\text{pressure drop (psi)} \times \text{flow(gpm)}}{1714 \times \text{pump efficiency}}$$

The most efficient hydraulic system is one that has no valves. However, such a system will also have very limited speed control. Many methods of control have been developed for hydraulic systems, and the method used depends on the types of pump and motor used and the characteristic of the load. Many of the systems are used on mobile equipment and machine tools, but they are not generally cost effective on industrial applications such as pumps and fans.

The Gibbs V/S drive shown in Fig. 7.31 is a packaged hydrostatic drive consisting of a constant-speed electric-drive

FIGURE 7.31 Gibbs V/S hydrostatic drive package. (Courtesy Gibbs Machine Co. Inc., Greensboro, NC.)

motor, a constant-speed hydraulic pump, and a variable-speed hydraulic motor. The hydraulic pump is a variable-volume positive-displacement pump, and the hydraulic motor is a fixed-volume positive-displacement motor. In the Gibbs package unit, the hydraulic pump is about 87% efficient over its working range, and the hydraulic motor has an efficiency of 92% over the working range. Figure 7.32 shows the hydraulic efficiency and the overall system efficiency for a 10-hp, 1800-rpm package unit operating over a 4:1 speed range, with a constant-torque load.

These types of adjustable-speed drives can operate from 0 to maximum speed at constant torque, with a recommended usable range of 27:1. The maximum output speed depends on the selection of the hydraulic motor in the package drive. These package drives

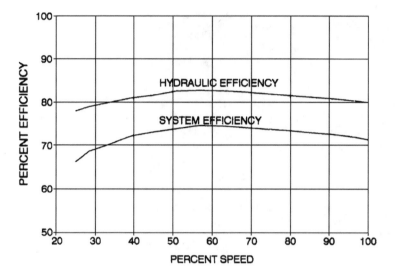

FIGURE 7.32 Efficiency of a hydrostatic package-drive unit driving a constant-torque load. (Courtesy Gibbs Machine Co., Greensboro, NC.)

are available up to 75 hp and can be provided with manual, electronic, pneumatic, or hydraulic controls. This type of drive is basically a constant-torque drive and is not normally used on variable-torque loads.

Hydroviscous Drives.   Another class of adjustable-speed fluid drives are the hydroviscous drive units. The basic components of the hydroviscous drive are (1) the torque-transmitting clutch plates, pressure plate, and flywheel assembly; (2) the oil pump for cooling and controlling oil; (3) the variable-orifice controller and control piston with a torque-limiting valve. Figure 7.33 is a cross section of one of these drives, manufactured by Great Lakes Hydraulic, Inc., showing the various components of the drive. Figure 7.34 shows a complete assembly for a horizontal unit.

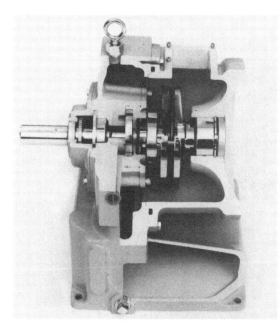

FIGURE 7.33 Cross section of a hydroviscous clutch assembly. (Courtesy Great Lakes Hydraulics, Inc., Grand Rapids, MI.)

There is a continuous flow of fluid between the constant-speed and adjustable-speed elements. The torque is transmitted through this film of fluid according to the oil shear principle. The amount of torque transmitted is proportional to the amount of piston pressure applied. As the piston pressure increases, the slip between the plates decreases. At the maximum rated piston pressure, the plates are locked in, and the output shaft is then running at input motor speed. The orifice controller determines the pressure supplied to the piston area. Minimum pressure is supplied to the piston when the orifice is completely open, bypassing fluid to the sump. The piston pressure is increased as the orifice is closed, and the slip between the clutch plates decreases. The orifice controller, which controls the piston pressure, can be manual, pneumatic, hydraulic, or electronic, as

FIGURE 7.34 Assembly of a hydroviscous drive package. (Courtesy Great Lakes Hydraulics, Inc., Grand Rapids, MI.)

required. With automatic control, the output speed can be regulated within ±2% of maximum speed. The torque transmitted by the drive is adjusted by changing the piston pressure.

This type of drive can be used for constant- and variable-torque applications and provides smooth operation at all speeds. The losses for these units include the slip loss that is common to all hydraulic drives and the fixed losses of the unit. Figure 7.35 illustrates the performance of a hydroviscous drive driving a vari-

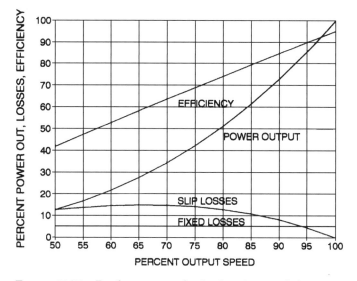

FIGURE 7.35 Performance of a hydroviscous drive system driving a variable-torque load. (Courtesy Great Lakes Hydraulics, Inc., Grand Rapids, MI.)

able-torque load such as a centrifugal pump or fan; these data do not include the drive motor losses.

7.2.6 AC Variable-Frequency Drives

The squirrel-cage induction motor is normally considered a constant-speed device with an operating speed 2–3% below its synchronous speed. However, efficient operation can be obtained at other speeds if the frequency of the power supply can be changed. The synchronous speed of an induction motor can be expressed by

$$N_s = \frac{120f}{p}$$

where

N_s = synchronous speed, rpm

f = power supply frequency, Hz

p = number of poles in motor stator winding

A four-pole induction motor that has a synchronous speed of 1800 rpm when operated on a 60-Hz power supply operates at the following synchronous speeds as the power supply frequency is changed:

Power frequency, Hz	Motor synchronous speed, rpm
120	3600
90	2700
60	1800
30	900
15	450
7.5	225

Variable-Frequency Power Supplies. The utilization of power semiconductor technology has provided an economic means to generate a variable-frequency power supply from a fixed-frequency power source for industrial applications. Using the output of this variable-frequency semiconductor power system to supply three-phase power to a three-phase induction motor provides a means to vary the speed of the induction motor. Today, these systems are commonly identified as adjustable-frequency controllers or adjustable-frequency drives. These "controllers" consist of two basic power sections: the converter section, which converts the incoming AC power to DC power, and the inverter section, which inverts the DC power to an adjustable-frequency, adjustable-voltage AC power.

The size and types of power semiconductors used in the power sections of the controller depend on the voltage level, power level, and type of inverter.

CONVERTER POWER SECTION. In the converter power section (AC to DC power), the power semiconductors are usually

1. *Silicon rectifiers.* These are commonly referred to as diodes. The silicon diode has the characteristic of permitting current flow in one direction and blocking current flow in the opposite direction. These rectifiers, along with the silicon control rectifiers (SCRs), are the workhorses of the semiconductors for power conversion. They range in current rating up to 4800 A and voltage rating up to 5000 V. These devices have no control characteristics and are either conducting or blocking power.

2. *Silicon control rectifiers or thyristors.* The silicon control rectifiers block current flow in one direction and permit current flow in the opposite direction, much as the silicon diode does. Unlike the diode, however, the start of current flow can be controlled in the SCR. The SCR switches on and conducts current from the anode to the cathode when a proper voltage pulse is applied to the gate terminal. Current continues to flow until the device switches itself off. The SCRs have large power handling capability. They range in current rating up to 4000 A and in voltage ratings up to 4500 V. The rating of the device depends on the case temperature and duty cycle of the application.

3. *Gate turn-off thyristors.* The gate turn-off thyristor (GTO) is a semiconductor device that can be turned on like the thyristor (SCR) with a single pulse of gate current, but it can also be turned off by the injection of a negative gate current pulse. The GTO power losses are higher during switching, but elimination of forced commutation circuits improves the overall efficiency of the converter. In addition, GTOs are suitable for higher switching frequencies than SCRs. They are available with turn-off current ratings up to 3000 A as well as a blocking voltage capability up to 4500 V.

INVERTER POWER SECTION.   In the inverter power section (DC
power to AC power), the power semiconductors used depend on
the type of inverter, voltage, and power ratings and may be any of
the following:

1. *Silicon control rectifiers.* See above comments on SCRs.
   Because of their limited switching frequency, these
   devices are generally not used in pulse width modulation inverters.
2. *Gate turn-off thyristors.* See above comments on GTOs.
   Again, the frequency of operation is limited but is
   higher than the switching frequency of the SCR. GTOs
   have been used in pulse width inverters.
3. *Bipolar transistors.* These devices can be switched at
   higher frequencies than SCRs. However, the current
   ratings are limited; they may be on the order of 400 A
   rating, with VCEO ratings of 600 V, and 120 A rating,
   with 1000-V VCEO. Significant drive power is required
   for these devices.
4. *Bipolar Darlingtons.* These devices are generally two-
   or three-stage devices with built-in emitter-base
   resistances, speed-up diodes, and freewheeling diodes.
   The frequency of operation is typically in the 5- to 8-
   kHz range, but the devices can operate at higher
   frequencies, and the gain is considerably higher than
   for the bipolar transistor. Bipolar Darlingtons are
   available in the range of 140 A at 1400 V and 600 A at
   1200 V. The units can be operated in parallel, and this
   is common practice in many inverters, with as many as
   four devices in parallel.
5. *Insulated gate bipolar transistors* (IGBTs). The IGBT
   combines on a single chip the high-impedance, voltage-
   controlled turn-on and turn-off capabilities of power
   MOSFETS and the low on-state conduction losses of
   the bipolar transistors. These devices can be switched
   at higher frequencies than the Darlington units and
   can be connected in parallel. They also have lower

base-power requirements than the Darlington units. The ratings range up to current rating I_c of 600 A and voltage VCES of 1200 V. IGBTs are finding increased use in pulse width modulation inverters. The devices can be operated in parallel.

Figure 7.36 illustrates the relative rating of some of these power semi-conductor devices. The Darlington transistors and the IGBTs can be switched at frequencies above the range of human hearing. In addition, they can be operated in parallel. The types of power semi-conductor devices used in a particular type of inverter can change as the quality, capacity, and cost of existing devices and new devices improve.

Types of AC Inverters. The AC three-phase induction motor can be used for adjustable-speed applications when the power to the motor is supplied by a variable-frequency power supply (inverter).

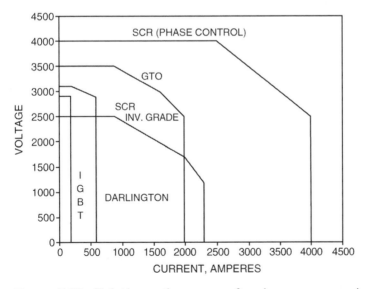

FIGURE 7.36 Relative rating range of various power semiconductor devices. (Courtesy Powerex, Inc., Youngwood, PA.)

The input voltage to the motor is varied proportionally to the frequency, i.e., at constant volts/hertz. At low frequencies, however, the voltage may be increased above its proportional level to obtain adequate torque. The torque developed by the induction motor is proportional to the magnetic flux in the motor air gap and to the rotor slip. As the frequency is decreased, the reactance of the motor decreases so that the applied voltage must be decreased proportionally to the frequency decrease to maintain constant air gap flux. If the applied voltage is not decreased, the motor magnetic circuit becomes saturated and there are excessive motor losses. At normal frequencies, the stator winding resistance drop is only a small percentage of the stator voltage drop so that the difference between the applied voltage and the net air gap voltage is relatively small. However, since the stator resistance is constant as the frequency is decreased and the reactance decreases proportionally to the frequency, the stator resistance drop voltage becomes a high percentage of the applied voltage. This results in a decrease in the net air gap voltage and air gap flux. Therefore, at lower frequencies

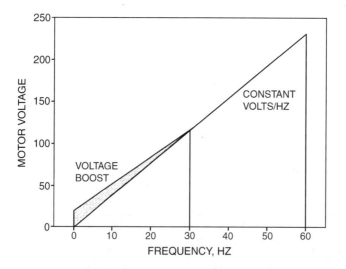

FIGURE 7.37 Typical voltage boost compared to constant volts/hertz motor voltage.

(about 10 Hz and lower), to compensate for this increased stator resistance voltage drop and maintain the flux in the air gap, the applied voltage must be increased above the constant volts/hertz level. Figure 7.37 shows the typical voltage boost compared to constant volts/hertz at the lower frequencies. The amount of voltage boost should be limited so that the current drawn by the motor does not exceed 150% of the current rating of the adjustable-frequency power supply. If a higher motor current is needed to achieve the necessary starting torque, a higher current–rated adjustable-frequency power supply will be required. This high-voltage boost should be maintained only during the starting of the motor to protect both the drive motor and the inverter from damage. A number of inverter types are used in adjustable-frequency power supplies, but the most common types are

    Voltage-source inverters
    Current-source inverters
    Pulse width modulation inverters
    Vector control inverter systems

VOLTAGE-SOURCE INVERTER. Figure 7.38 illustrates the basic power circuit for a variable-voltage-source, six-step inverter. In this system, the 60-Hz input voltage is converted to a DC adjustable voltage by means of a three-phase semibridge converter. Then, by

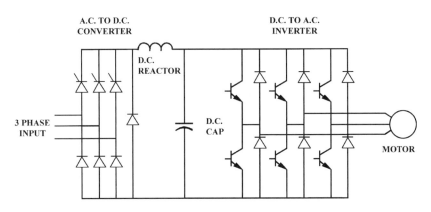

FIGURE 7.38 Voltage-source inverter.

means of a DC-to-AC transistor inverter, each of the three-phase output lines is switched from positive to negative for 180° of each 360° cycle. The phases are sequentially switched at 120° intervals, thus creating the six-step line-to-neutral voltage, or square wave line-to-line voltage, as shown in Fig. 7.39. The DC power supply is normally controlled by SCRs in the bridge rectifier, and an LC filter is used to establish a stiff DC voltage source. The output frequency is controlled by a reference signal that sets the control logic to achieve the correct gate or base-drive signals for the semiconductors in the inverter section. The semiconductors in the inverter section can be SCRs, GTOs, transistors, or Darlington transistors.

Speed control beyond the 10:1 range becomes a problem with the six-step inverter because at low voltage and frequency the harmonic currents become excessive, causing motor heating, torque pulsations, and cogging.

Advantage of the voltage-source inverter include

- Inverter section can use SCRs, GTOs, or transistors.
- Low switching frequency devices can be used.
- It is the simplest regulator.
- Standard or energy-efficient motors can be used with proper derating.

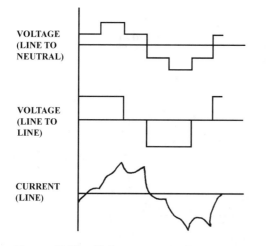

FIGURE 7.39 Voltage-source inverter wave shapes.

- It has multimotor capability.
- Voltage stress on motor insulation system is low.

Disadvantages of the voltage-source inverter include

- Poor input power factor that decreases with decreasing output frequency
- Harmonics fed into the 60-Hz AC supply system
- Limited speed control beyond the 10:1 range
- Torque pulsations and cogging
- High-harmonic currents, causing excessive motor heating

CURRENT-SOURCE INVERTER. In contrast to a stiff voltage source as in a voltage-source inverter, the current-source inverter has a stiff DC current source at the input. This is generally accomplished by connecting a strong inductive DC filter reactor in series with the DC source and controlling the voltage within a current loop. Figure 7.40 illustrates the power circuit for the current-source inverter. A three-phase bridge consisting of six SCRs converts the AC input to DC, and a three-phase bridge autosequential-commutated inverter inverts the DC to the AC output voltages. With a stiff current source, the output current waves are not affected by the load. The power semiconductors in the

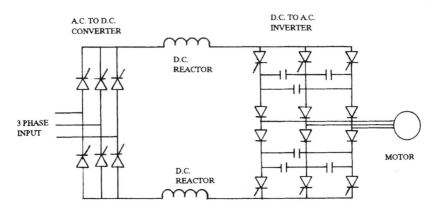

FIGURE 7.40 Current-source inverter.

current-source inverter have to withstand reverse voltages; therefore, devices such as transistors and power MOSs are not suitable. Figure 7.41 illustrates the waveforms for the current-source inverter. Note the high spikes in the line-to-neutral voltage.

Advantages of the current-source inverter include

- Simples SCR-type circuit
- Low-frequency inverter switches
- Inherent short-circuit capability
- Inherent regeneration capability
- Rugged construction

Disadvantages of the current-source inverter include

- Motor and control must be matched.
- Not suitable for multimotor operation.
- Poor input power factor.
- Low-speed torque pulsations and cogging.
- It can cause high-voltage spikes at the motor.

Current-source inverters have been developed using GTO devices and pulse width modulation to overcome some of the disadvantages of the current-source inverter.

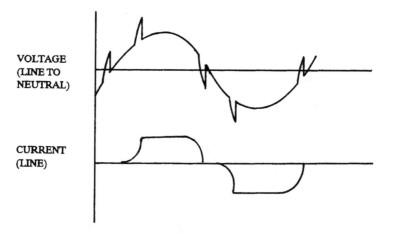

FIGURE 7.41   Current-source inverter wave shapes.

PULSE WIDTH MODULATION INVERTERS. In the pulse width modulation (PWM) system, the input AC power is rectified to a constant potential DC voltage. The DC voltage is then applied to the motor in a series of pulses. A number of methods have been devised to control the pulse width and to vary the frequency of the pulses as the motor speed is changed. In some cases, at the higher speed, the system becomes a six-step inverter.

Figure 7.42 shows the power circuit for a PWM inverter with a diode bridge to convert the AC voltage to DC voltage and a transistor DC to AC inverter to generate the AC output voltage.

The technology of pulse width modulation is not new. However, the use of microprocessors to provide improved modulation techniques and higher-speed switching power semiconductor devices such as transistors and IGBTs are making the PWM inverter the standard inverter in the 1- to 500-hp range.

A number of pulse width modulation procedures are used in today's PWM inverters. Some of these are

- Sinusoidal with a sine wave signal and a triangular carrier wave
- Harmonic elimination, particularly the fifth, seventh, eleventh, and thirteenth harmonics

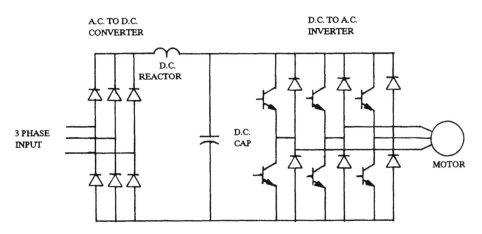

FIGURE 7.42 Pulse width modulation inverter.

- Distortion minimization with five switching angles/ quarter cycle
- Minimum ripple current
- Uniform sampling

The ideal PWM system balances the switching losses in the inverter with the current and torque ripples and the heating losses in the drive motor for the best overall performance (Figure 7.43).

By selecting the width and spacing of the pulses, lower-order harmonics, such as the fifth, seventh, and eleventh, can be eliminated in the waveform. If the pulse rate is high enough, the motor inductance presents a high impedance so that the pulse-rate-frequency current is insignificant. From the motor viewpoint, it is desirable to have a high-frequency pulse rate. From the inverter viewpoint, since most of the losses occur during switching, it is best

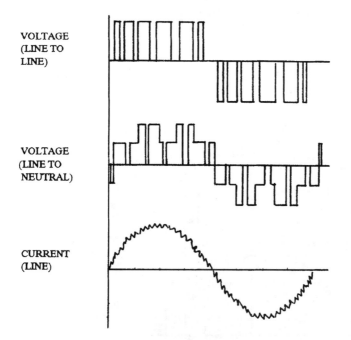

Figure 7.43   Pulse width modulation inverter wave shapes.

to have a low pulse rate. However, the number of pulses per cycle must be maintained high enough to avoid troublesome harmonics that may be resonant with the motor components and cause noise and vibration in the drive. The switching and recovery time of SCRs limits their use on PWM systems. IGBTs, power transistors, Darlington transistors, and GTOs have faster switching times with lower losses, and so they are used at the higher pulse rates required for smooth operation. While the PWM inverters improve the waveforms by eliminating the low-order harmonics, they impose a series of high-voltage impulses on the motor winding. Although the winding inductance smooths the current waveform, the rapid voltage changes produce insulation stresses on the first few turns of each of the motor windings. Full-voltage PWM systems produce the most severe stresses, particularly at low speeds, where the motor back-EMF is low.

Advantages of PWM inverters include

- Wide speed range.
- Smooth low-speed operation.
- Multimotor operation.
- Standard or energy-efficient motors can be used with proper derating.
- Minimum problems matching motor and inverter.
- High-input power factor.

Disadvantages of PWM inverters include

- Complex control
- Requires high-frequency power semiconductors in the inverter
- Higher motor heating and noise (depends on the modulation system used)
- Not regenerative
- Imposes high-voltage gradients on the motor insulation system

There are numerous variations of these three types of adjustable-frequency inverters, but the principle of operation is essentially the same. As with any product, changes and improvements are being accomplished every day. The improvements come primarily from the increasing use of integrated circuits as well as microprocessors, which have greatly reduced the number of control logic components. Also the use of power IGBTs, power transistors, and GTOs has reduced the cost of power elements. These factors, plus improved designs and techniques, have reduced and will continue to reduce the size and cost of the inverters. At the same time, performance and reliability continue to improve.

VECTOR CONTROL INVERTER SYSTEMS. One important variation or addition to the previously discussed inverter systems is the vector control inverter system. Vector control considers the analogy between AC and DC electrical machines. The ultimate object of the vector control system is to control the AC induction motor as a separately excited DC motor is controlled, i.e., to control the field excitation and torque-generating currents separately and independently. To control the induction motor in this manner, the air gap flux (net air gap voltage) and rotor current must be separately controlled. The vector control drives available are based mostly on the indirect flux control method. The magnitude, frequency, and phase of the stator current components are controlled as a function of the rotor position, slip frequency, and torque command. Figure 7.44 is a block diagram of a vector control

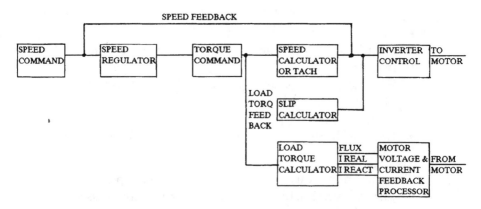

FIGURE 7.44 Block diagram of vector control logic.

logic system, with the input signals received from the motor and the output control signal from the vector control system supplied to the inverter section of a PWM inverter. The excitation current component and the torque component of the current are calculated from the motor terminal voltage and current and the motor speed. The performance of this method of control depends on how closely the algorithm of the vector system matches the induction motor characteristics. The precision of the system also depends on the precision of the tachometer or rotor speed sensor since the slip control is based on this signal. Without a tachometer or. rotor speed sensor, the precise speed range is 20:1, with the speed

FIGURE 7.45 Family of PWM adjustable-frequency power supplies. (Courtesy Magnetek, New Berlin, WI.)

within ±0.5% accuracy and an allowable operating range of 50:1. With a precision rotor speed or position sensor, the operating range can be extended to 200:1, with a speed within ±0.1% accuracy. Closer speed regulation can be obtained with more precise rotor sensing. This system can be used with either voltage-source or current-source inverters with a PWM inverter section.

These systems have been used mainly on servosystems and machine tool applications, but are being used in more industrial application to replace DC adjustable-speed drives.

Inverter Features.   The available range of features and rating of adjustable-frequency drives has expanded extensively in recent years. Figure 7.45 shows a family of Magnetek adjustable-frequency systems. These units are PWM units with an insulated-gate bipolar transistor in the inverter section. The range in rating for these units is 1–40 hp at 230-V input and 1–75 hp at 460-V input. Units are also available with transistors in the inverter section in the range of 1–125 hp at 230-V input and 1–600-hp input at 460-V input.

The control and protection features available for this product line, which is typical of many adjustable-frequency systems, are shown in Fig. 7.46.

In addition to the usual protection features, a variety of adjustments are available, including the following:

- Adjustable acceleration and deceleration rates.
- Torque limit control.
- Stall protection.
- Critical frequency rejection. Generally, three prohibited frequencies can be selected to prevent the drive system from operating at a resonant speed at speeds within the operating speed range.
- Selectable volts/hertz pattern. Normally, the output of the inverter section is based on constant volts/hertz. However, to provide for application variations, such as induction motor rating, high starting torque, variable-torque loads, and operation at different frequency ranges, provision is made for different volts/hertz patterns. Figure 7.47 illustrates the types of volts/hertz

STANDARD FEATURES

- Frequency resolution 0.1 Hz with digital reference; 0.06 Hz with analog reference
- Frequency regulations 0.01% with digital command; 0.1% with analog (15 to 35° C)
- Standard Frequency range 1.5 to 800 HZ
- Volts/Hertz ratio, 15 preset patterns, one fully adjustable pattern
- Independent accel/decel 0.1–6000 sec.
- DC injection braking amplitude & duration, current limited
- Signal follower external bias & gain
- Critical frequency rejection, 3 selectable
- Torque limit, 30–150%
- Jog speed, adjustable zero to 100%
- Multi-speed setting, 9 possible
- Forward/reverse operation
- Speed range 40:1
- NEMA 1 enclosure (NEMA 12 optional)

- 24 VDC Logic
- Run/fault contacts 1 amp, 250 VAC or 30 VDC
- Remote speed reference capability 0–10 VDC (20K ohms) or 4–20 mA (250 ohm)

PROTECTION and MONITORING

- Overload capability to 150% rated, 60 sec.
- Instantaneous overcurrent trip and indication
- Overvoltage trip and indication
- Undervoltage trip and indication
- Overtemperature trip and indication
- External fault trip and indication
- Blown fuse trip and indication
- Control circuit error trip and indication
- DC bus charge indication
- DC bus fuse
- Ground fault protection
- Stall prevention
- Electronic motor overload protection

- Momentary power failure ride-through (2 sec. 5 HP and above); (0.2 sec. below 5 HP with option of 2 sec.)

ENVIRONMENTAL CONDITIONS

- Altitude to 3300 feet above sea level
- Operating ambient temperature −10 to 40° C
- Storage temperature −20 to 60° C
- Noncondensing relative humidity to 90%
- Vibration 1G max under 20 Hz; 0.2G at 20–50 Hz

INPUT POWER REQUIREMENTS

- 230V model for 200, 208, 220 or 230 VAC, ±10%
- 460 V model for 380, 400, 415, 440, or 460 VAC, +10%
- 3-phase, 3-wire, phase sequence insensitive
- Frequency 50 or 60 Hz, ±5%

FIGURE 7.46 Typical features of PWM adjustable-frequency power supplies. (Courtesy Magnetek, New Berlin, WI.)

patterns that can be selected. In addition, some units automatically select the optimum voltage for a given frequency and load condition.

- Automatic carrier frequency. As the motor load increases, or the operating frequency decreases, the carrier frequency will automatically increase; this increase in the carrier switching frequency reduces the output current harmonics and, as a result, provides more motor torque per ampere.

7.2.7 AC Variable-Frequency Drive Application Guide

Unfortunately, the selection and application of an AC variable-frequency induction motor drive system are more complex than the

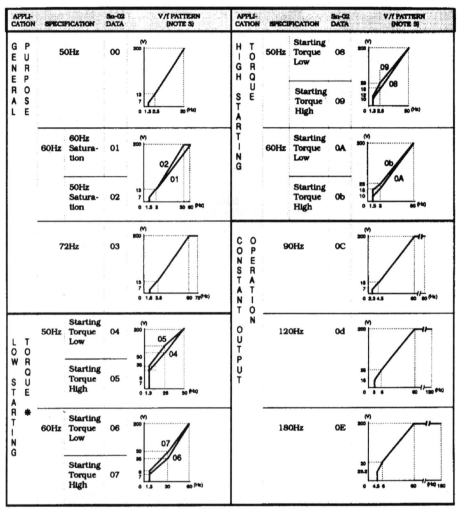

NOTES: *Use of these patterns is NOT RECOMMENDED. Consult MagneTek for assistance.
1. The following conditions must be considered when selecting a V/f pattern: —Pattern matches the voltage-frequency characteristics of the motor.—Maximum motor speed.
2. V/f pattern for high starting torque should be selected for:—Wiring distance.—Large voltage drop at start.—AC reactor connected to GPD 503 input or output.—Use of motor rated below GPD 503 max. output.
3. Patterns shown are for 200 V input; for other input, multiply all V values by $V_{IN}/200$. Example: For 400 V input, multiply by $400/200 = 2$.

FIGURE 7.47 Typical standard (preset) volts/hertz patterns. (Courtesy Magnetek, New Berlin WI.)

selection of a fixed-speed induction motor. The duty cycle of the inverter-motor combination must be checked at all load conditions to make certain that the particular drive combination is suitable for the given application. In addition, some applications may require control options such as digital speed control, closed-loop speed control, frequency metering, variable-voltage boost, or remote signal inputs such as pressure or temperature. The characteristics that determine the appropriate drive combination are the following:

1. Speed range required
2. Speed-torque characteristics of the load
3. Load inertia
4. Load acceleration and deceleration times
5. Required operating time at various speeds
6. Inverter output waveform and its approximate harmonic content
7. System efficiency over the operating range
8. Regenerative energy dissipation in the inverter
9. Motor temperature rise at the required duty cycle and the voltage-to-frequency ratio provided by the inverter
10. Motor rating based on the duty cycle
11. Motor insulation life derating for its input waveform (applies to full-voltage PWM systems)
12. Inverter construction and enclosure
13. Motor enclosure

When applying adjustable-frequency induction motor systems, the characteristics of both the motor and the power supply must be considered as an integrated system. The constraints imposed on the induction motor must be considered in selecting both the motor and the inverter.

The adjustable-frequency power supply has a constant ratio of volts/hertz. As the output frequency and voltage are changed, the induction motor speed changes proportionally to the output frequency. Since the induction motor reactance is proportional to the frequency, the output voltage must also decrease in the same ratio to maintain a constant air gap flux. If a constant slip rpm is maintained, the induction motor is essentially a constant-torque motor. For example, consider a four-pole induction motor that, when supplied with 60-Hz power, produces its rated torque at 1750 rpm, or a slip of 50 rpm. When the motor is supplied with 30-Hz power at constant volts/hertz, it will produce rated torque at 50-rpm slip, or $900 - 50 = 850$-rpm output speed. When the motor is supplied with 10-Hz power at constant volts/hertz, it will again produce rated torque at 50-rpm slip, or $300 - 50 = 250$-rpm output speed.

A conventional speed-torque curve exists for the induction motor at each frequency it operates at, as shown in Fig. 5.13. However, because of the 150% current limitation of the power supply, the maximum torque available is reduced as shown in Fig. 5.14. This reduced available torque and the time constraint of the 150% current limitation (usually for 60 sec) must be considered in the acceleration and deceleration of the load. This is particularly important for high-inertia loads. The locked-rotor or starting torque available is also limited by the current constraints of the power supply and the characteristics of the induction motor. At low frequencies below 10 Hz, voltage boost above the constant volts/hertz level may be required to start the load. This increase in voltage can cause excessive currents in both the motor and power supply and should be maintained for a minimum time. Figure 5.15 shows the locked-torque characteristics for a 10-hp motor at low frequencies, including the effect of voltage boost on the locked-rotor torque. The determination of the starting torque and accelerating torque are essential for the proper selection of the adjustable-frequency power supply and the drive motor.

These restrictions are often overlooked in the conversion from a fixed-speed drive to an adjustable-frequency motor drive system. Similarly, in the conversion from a mechanical variable-speed drive to an adjustable-frequency induction motor drive, the torque requirements must be determined. In many instances, the torques the mechanical drive can develop at low speeds and at starting exceed the torques available on the same rated horsepower adjustable-frequency drive.

Once the torque requirements have been established, the sizing of the induction motor must be based on the limiting temperature rise of the motor. The items that influence the temperature rise are operating speed range, type of load, motor losses, and type of motor enclosure. The induction motor losses, including the harmonic losses when the motor is operating from an adjustable-frequency power supply, must be balanced against the heat-dissipation ability of the motor. Figures 5.18 and 5.19 illustrate the increase in the induction motor losses at constant torque when supplied by a nonsinusoidal power source. In contrast, Fig. 5.21 shows the decrease in the heat-dissipation ability of the induction motor as the operating frequency is decreased. Figure 5.22 shows the influence of the harmonic losses on the temperature rise of a 5-hp induction motor at various operating frequencies.

Most manufacturers of electric motors have guidelines on the heat-dissipation capabilities of their motors at various speeds and types of loads.

Chapter 5 suggests some guidelines on the derating of motors for various types of applications when supplied by adjustable-frequency power systems.

Most adjustable-frequency power systems are provided with means to adjust the volts/hertz ratio and to provide a voltage boost when required. However, the lowest possible volts/hertz setting for satisfactory system operation should be selected. Unnecessarily high volts/hertz settings reduce the induction motor efficiency, increase the losses, and increase the motor temperature rise and noise level. Some adjustable-frequency power systems automatically adjust the voltage to the optimum level of motor operation. Figure 7.48 shows a comparison of the efficiencies for a 10-hp induction motor driving a variable-torque load at constant volts/hertz and at the optimum voltage for each frequency. In this case, the figure also shows the optimum voltage as a percent of the constant volts at each frequency of operation. Note that the voltage was maintained at constant volts/hertz down to about 45 Hz since there was no improvement in performance obtained by decreasing the voltage below constant volts/hertz.

If energy saving is the main justification for a drive, then the overall efficiency must be considered, including both the motor and adjustable-frequency power supply. The inverter efficiency varies

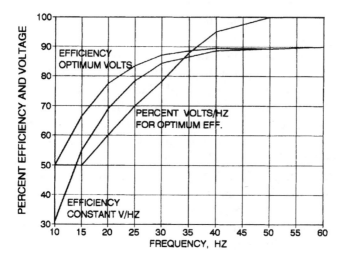

FIGURE 7.48 Improvement in the efficiency of a 10-hp, energy-efficient, four-pole induction motor by reducing the volts/hertz when supplying a variable-torque load.

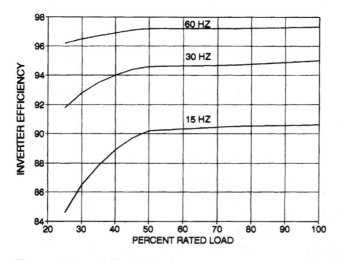

FIGURE 7.49 Adjustable-frequency power supply efficiency as a function of load and output frequency.

with the load and operating frequency. The motor efficiency varies with the load and the operating frequency, and there are additional losses in the motor as a result of the harmonics in the motor supply frequency. Figure 7.49 shows the efficiency of an adjustable-frequency power supply as a function of load and operating frequency. Figure 5.19 compares the efficiency of a 100-hp induction motor with a sinusoidal power supply and a nonsinusoidal power supply (such as an adjustable-frequency power source), reflecting the decrease in motor efficiency as a result of the harmonics in the supply voltage. The overall efficiency of the adjustable-frequency induction motor system is the product of the component efficiencies and is illustrated in Fig. 7.50. This figure also compares the induction motor efficiency when the motor is operating on a sine-wave power source to the overall efficiency when it is operating with an adjustable-frequency power system.

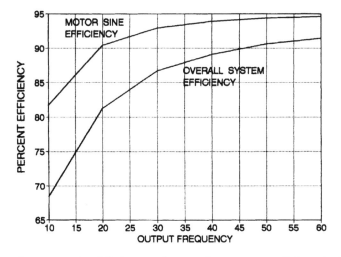

FIGURE 7.50 Efficiency of a 100-hp, energy-efficient induction motor with a constant-torque load on a sine-wave power supply versus the overall efficiency on an adjustable-frequency power supply.

The following is a summary of the types of loads suitable for application of adjustable-frequency induction motor systems:

- *Variable-torque loads*

  Centrifugal fans
  Centrifugal pumps
  Agitators
  Axial centrifugal compressors
  Centrifugal blowers

- *Constant-torque loads*

  Calenders
  Positive-displacement blowers
  Conveyers
  Centrifuges
  Reciprocating and rotary compressors
  Positive-displacement pumps
  Slurry pumps
  Cranes
  Elevators
  Mixers
  Printing presses
  Washers

- *Constant-horsepower loads*

  Drill presses
  Grinders
  Lathes
  Milling machines
  Tension drives
  Winders
  Recoilers

- *Impact loads.* The following types of impact loads may be suitable for application of adjustable-frequency induction motor systems but require special consideration of the adjustable-frequency power supply in order to provide the peak induction motor output torques required and stay within the current limitations of the adjustable-frequency power supply.

Lathes
Milling machines
Rolling mills
Punch presses
Shakers
Shears
Crushers

7.2.8 Wound-Rotor Motor Drives with Slip Loss Recovery (Static Kramer Drives)

The wound-rotor motor has normally been used for short-time duty applications such as cranes and hoists where torque control is of prime importance. When it has been used on continuous-duty installation, the major purpose has been to obtain controlled starting and acceleration. The reason for this limited use has been that high slip losses occur at speeds below normal operating speed. With the development of power electronics and solid-state inverters, systems have been developed to recover these slip losses.

These drives are commonly referred to as static Kramer drives. The original Kramer drives used a rotary converter instead of power semi-conductors and fed the power back to the line from a DC motor coupled to the induction motor. With the recovery of the rotor slip losses, the efficiency of the wound-rotor feedback system is comparable to the efficiency of an adjustable-frequency induction motor drive. It has an advantage in that the inverter has only to be large enough to handle the rotor slip losses. The system has the disadvantages, however, of the unavailability and high cost of the wound-rotor motor. Today, these systems are generally custom-designed for specific applications with a limited speed range, such as large pumps and compressors.

Figure 7.51 is a power circuit diagram for the static Kramer drive. As shown, the output of the wound rotor is connected to a three-phase rectifier bridge. The output of the bridge is connected to

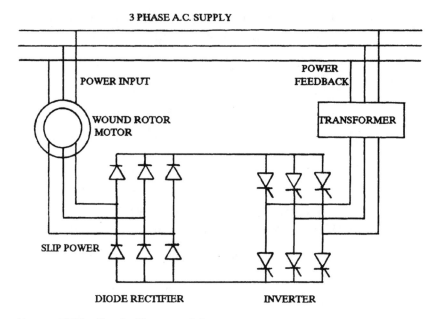

FIGURE 7.51 Static Kramer drive.

a fixed-frequency inverter, the output of which is connected to the primary power supply that supplies the motor stator. The connection from the inverter output to the primary power supply is generally through a matching transformer. The effective rotor resistance, and hence the motor speed, is controlled by controlling the firing angle of the power SCRs in the inverter section.

The speed range that can be obtained is determined by the motor secondary (rotor) voltage; for instance, for a 100% speed range system,

> 480-V power supply: The rotor voltage must be 380 V or less.
> 600-V power supply: The rotor voltage must be 480 V or less.

and for a 50% speed range system,

> 480-V power supply: The rotor voltage must be between 600 and 760 V.

600-V power supply: The rotor voltage must be between 750 and 960 V.

For power supply voltages above 600 V, such as 2300 and 4160 V, the motor primary can utilize the line voltage. However, a matching transformer is required at the output of the rotor inverter, and it need only be large enough to handle the rotor losses, not the total motor input.

The efficiency of the controller is approximately 98.5% and is constant over the speed range; thus, the system is very efficient in recovering the slip losses and raising the system efficiency.

Consider a 200-hp wound-rotor motor on a pumping installation where the motor horsepower load is a cubic function of the speed. Without the slip recovery controller, at full speed (1764 rpm):

Horsepower output, 200 hp
Motor efficiency, 94%

At one-half speed (882 rpm):

Horsepower output, 25 hp
Motor efficiency, 46%

With the slip recovery controller, at full speed (1764 rpm):

Horsepower output, 200 hp
Motor efficiency, 94%
Overall system efficiency, 94%

At one-half speed (882 rpm):

Horsepower output, 25 hp
Motor efficiency, 46%
Overall system efficiency, 84%

Note that recovery of the slip losses at one-half speed increased the efficiency from 46 to 84%.

The wound-rotor motor with a slip recovery system used on a pump or fan application can usually be operated over a 50% speed range with self-ventilation. For applications requiring continuous operation below 50% speed, forced ventilation may be required for the motor.

This type of drive system offers an energy-efficient system comparable to adjustable-frequency systems and superior to slip-loss systems.

7.3 APPLICATIONS TO FANS

Various types of fans used to move air or other gases are among the largest consumers of electric power and among the largest users of integral-horsepower electric motors. In general, fans can be divided into two broad categories: centrifugal fans and axial flow fans.

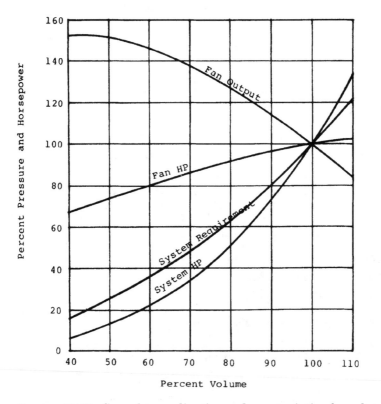

FIGURE 7.52 Sample application, characteristic fan data, and system characteristic.

The guide for selection of the proper fan for a given application is covered by most fan manufacturer catalog information. However, to obtain the most energy-efficient system, it is necessary to examine the methods of controlling airflow in a given system.

First, it is necessary to determine the system resistance characteristic for various airflow rates. A curve can be developed for the system in terms of the air volume versus the static pressure required. This curve generally follows a simple parabolic law in which the static pressure or resistance to airflow varies as the square of the volume of air required. Figure 7.52 shows the system curve for a specific example that will be discussed later.

Next is the selection of the type of fan. Many of the applications in general heating, ventilating, and air conditioning systems involve centrifugal fans. These fans generally fall into three major categories based on the type of impeller design:

1. *Backward-curved blades.* The horsepower reaches a maximum near peak efficiency and becomes lower toward free delivery. Figure 7.53 is the typical performance curve for the backward-curved fan. The volume is the percent of free-flow volume, and the pressure is the percent of static pressure at zero volume. The horsepower is the percent of maximum horsepower.
2. *Radial blades.* These have higher-pressure characteristics than the backward-curved fan. The horsepower rises continually to free delivery. Figure 7.54 is the typical performance curve for radial-blade fans.
3. *Forward-curved blades.* The pressure curve is less steep than that for the backward-curved fan. The peak efficiency is to the right of the peak pressure. The horsepower rises continually to free delivery. Figure 7.55 is the typical performance curve for forward-curved fans.

The size and type of fan selected should be such that the fan is operating near its peak static efficiency for the maximum flow rate

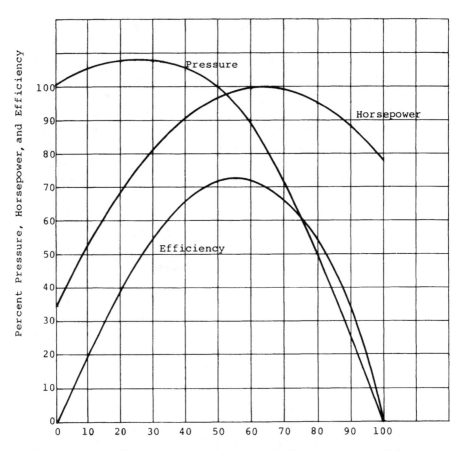

FIGURE 7.53 Characteristic curves for the backward-curved fan.

required. The performance of the fan at other speeds will follow the following fan laws:

1. The volume V of air varies as the fan speed.
2. The static pressure P varies as the square of the fan speed.
3. The horsepower varies as the cube of the fan speed.

$$\text{Static efficiency} = \frac{VP}{6369 \times \text{hp}}$$

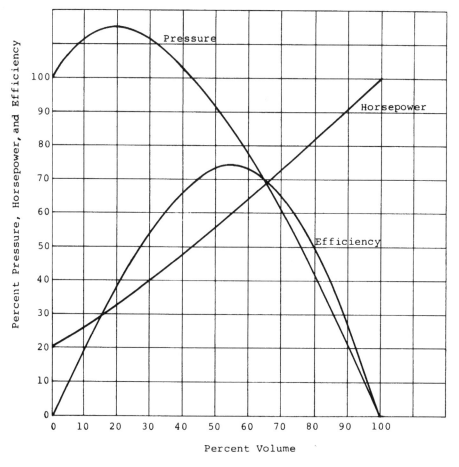

FIGURE 7.54 Characteristic curves for the radial fan.

Several methods of controlling the airflow can be considered:

Damper control
Variable inlet vane control
Hydrokinetic or fluid-drive systems
Eddy current drive systems
Mechanical variable-speed drive units
AC variable-frequency systems
Wound-rotor motor with slip recovery systems
Two-winding, two-speed motors

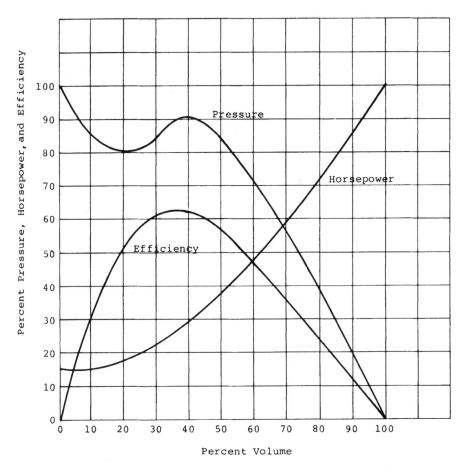

FIGURE 7.55 Characteristic curves for the forward-curved fan.

The comparison of these systems is best illustrated by an example.

EXAMPLE

The characteristic curves of the fan and air system for the example are shown in Fig. 7.52. The data have been shown in percentages of the pressure, volume, and horsepower at the balanced point at which the system resistance curve and the fan curve intersect.

Based on this fan-system curve, the horsepower input at the fan to produce the airflow is also shown. Figure 7.56 shows the input horsepower required for each of the preceding systems compared to the fan horsepower required as the airflow requirement is varied between 40 and 100% volume.

To determine the most energy-efficient system, the operating cycle, i.e., the percent time operating at each volume flow, must be determined; then the energy savings and net present worth of each system can be compared to the most inefficient system, i.e., one using a discharge damper.

To illustrate these data, three different operating cycles have been assumed with the following data:

Horsepower at full air volume, 100 hp
Annual operating hours, 4000 hr
Initial power rate, $0.06/kWh
Annual increase in power rate, 10%
Cost of money, 15%
Tax rate, 40%
System life, 10 yr

The operating cycles are as follows:

| Cycle | Time at each air volume, % | | |
	100% air vol.	80% air vol.	60% air vol.
I	75	15	10
II	33	33	34
III	10	15	75

Two other operating cycles based on fixed-speed operation at two speeds using a two-winding, two-speed motor with an 1800/1200-rpm speed combination are as follows:

Operating cycle 1-A:

100% (full) air volume: 75% of the time
66% air volume: 25% of the time

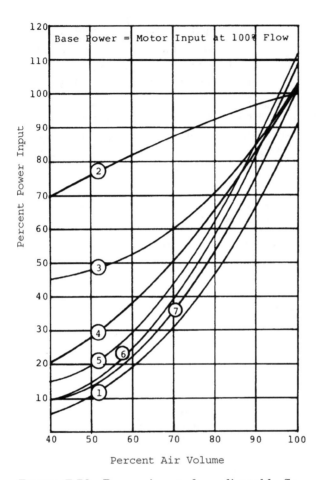

FIGURE 7.56 Power input for adjustable-flow systems: (1) fan horsepower required, (2) damper control, (3) inlet vane control, (4) hydrokinetic and eddy current system, (5) mechanical varispeed system, (6) variable-frequency system, (7) wound rotor with rotor loss recovery system; 100% power is the power input to the motor at 100% airflow.

TABLE 7.2 Comparison of Annual Kilowatt-Hour Savings for Adjustable Flow Fan Systems[a]

Operating cycle	System					
	Multispeed motor	Variable vane inlet	Fluid drive or eddy current drive	Mechanical varidrive	Variable-frequency system	Wound rotor with slip recovery
I	—	10,431	15,080	4,070	12,550	32,768
II	—	51,515	71,916	81,355	87,879	103,008
III	—	80,382	118,948	141,854	148,216	157,834
I-A	39,396	—	—	—	—	—
III-A	129,060	—	—	—	—	—

[a] Base for comparison: damper control.

Operating cycle 1-B:

100% (full) air volume: 25% of the time
66% air volume: 75% of the time

The summary of the annual power savings in kilowatt-hours for each system is shown in Table 7.2. The net present worth of the energy savings based on the defined assumptions is shown in Table 7.3. This summary shows the influence of the operating cycle on the power savings and the net present worth of those savings. When compared to the first cost of each system in conjunction with other considerations, including reliability, flexibility, maintenance, and environment, this net present worth will provide an economic basis for selecting the most cost-effective system.

7.4 APPLICATIONS TO PUMPS

Pumps are the largest user of electric motors in the integral-horsepower sizes 1 hp and larger. The selection and application of electric motors and adjustable-speed systems to pumps become very complex and difficult because of the large numbers of types of pumps and their lack of standardization. The pumps fall into two broad categories: displacement pumps and dynamic pumps. The dynamic pumps include noncentrifugal and centrifugal types. The highest percentage of the pumps used for industrial processes are of the centrifugal type; therefore, the discussion in this section will be limited to drive applications of this type.

In the selection of a pump for a given system, the system characteristics must be determined in terms of flow rate in gallons per minute versus total head, in feet, under all flow rates expected. The system should include an allowance for pipe corrosion and other factors that affect the system characteristics. The pump then selected should be sized to the system characteristic at the maximum flow rate such that the efficiency is close to the optimum for the pump. Then, for an adjustable-speed system, it is necessary to check the pump performance at the minimum flow point required.

The various pump manufacturers provide data for their pumps and also provide assistance in selecting the correct pump for the

TABLE 7.3 Net Present Worth Comparison in Dollars for Adjustable-Flow Fan Systems[a]

Operating cycle		System				
	Multispeed motor	Variable vane inlet	Fluid drive or eddy current drive	Mechanical varidrive	Variable-frequency system	Wound-rotor motor with rotor loss recovery
I	—	3,558	5,143	1,388	4,271	10,176
II	—	17,571	24,528	27,746	29,973	35,029
III	—	27,416	40,569	48,381	50,551	53,831
I-A	13,727	—	—	—	—	—
III-A	44,018	—	—	—	—	—

[a] Base for comparison: damper control.

application. The initial guide is a performance curve at a fixed speed, as shown in Fig. 7.57, indicating the best efficiency point on each pump in a family of pumps. For a given pump, detail performance curves such as those in Fig. 7.58 are available from the manufacturer.

The following relationship applies to pumps (as discussed for fans in Sec. 7.3):

$$\frac{Q_1}{Q_1} = \frac{N_1}{N_2}$$

$$\frac{H_1}{H_2} = \left(\frac{N_1}{N_2}\right)^2$$

$$\frac{BHP_1}{BHP_2} = \left(\frac{N_1}{N_2}\right)^3$$

$$BHP = \frac{QH \times \text{Sp. Gr.}}{3960 \times \text{Pump Eff}}$$

$$\text{Pump Eff} = \frac{QH \times \text{Sp. Gr.}}{3960 \times BHP}$$

where

Q = capacity, gpm
H = total head, ft
BHP = brake horsepower
N = pump speed, rpm
Sp. Gr. = specific gravity of liquid

To illustrate the application and cost analysis of various methods of flow control, consider an example comparing the following:

Throttling in the discharge line
Eddy current or hydraulic fluid drive systems
AC variable-frequency system
Wound-rotor motor with slip recovery system

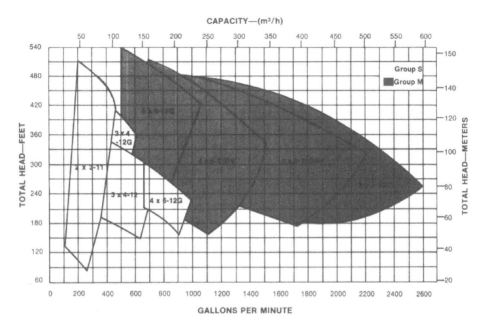

1780 R.P.M. Performance Curve

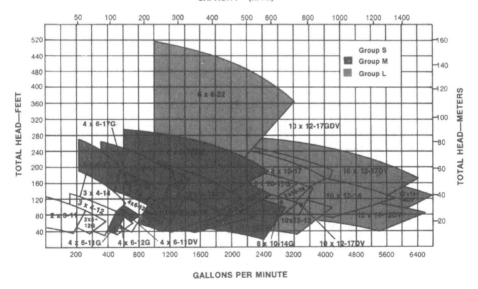

FIGURE 7.57 3550- and 1780-rpm centrifugal pump performance curves. (Courtesy Gould's Pumps Inc., Seneca Falls, NY.)

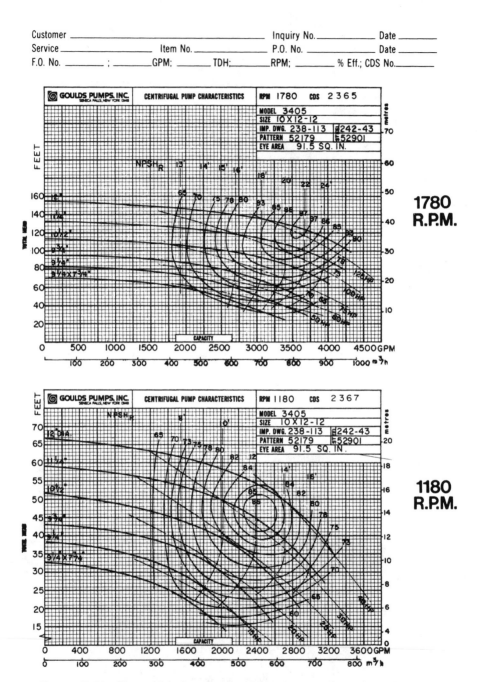

FIGURE 7.58 Centrifugal pump performance at the selected speeds 1780 and 1180 rpm. (Courtesy Gould's Pumps Inc., Seneca Falls, NY.)

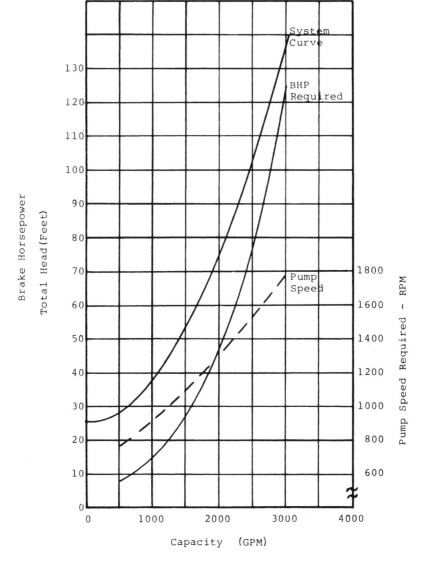

FIGURE 7.59 Sample fluid system characteristic.

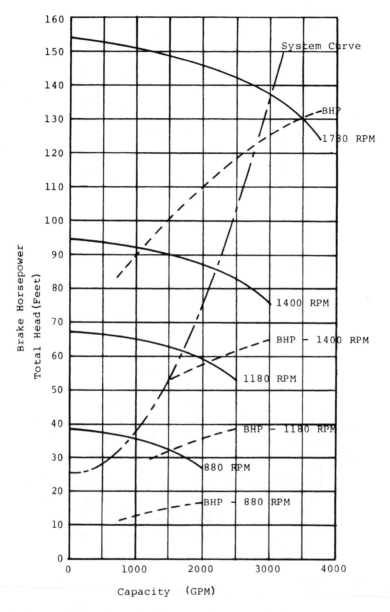

FIGURE 7.60 Centrifugal pump characteristics.

TABLE 7.4 Summary of Sample Calculations for Adjustable-Flow Systems

Flow rate, %	Flow, gpm	Pump[a] BHP required	Throttling system		Eddy current drive or fluid drive		AC variable-frequency system		Wound-rotor motor with slip loss recovery	
			Input, kW	Losses, kW	Input, kW	Losses, kW	Input, kW	Losses, kW	Input, kW	Losses, kW
100	3000	125	100.81	7.56	107.26	14.0	103.96	10.71	101.91	8.66
75	2250	60	92.65	47.89	61.29	16.5	51.15	6.39	51.45	6.69
50	1500	28	80.56	59.67	36.89	16.0	24.60	3.72	25.17	4.28

[a] For adjustable-speed pump.

Figure 7.59 shows the system flow characteristic and the centrifugal pump characteristics at various pump speeds for the specific application. Figure 7.60 shows the pump characteristic at various speeds and the system characteristics as a function of flow requirements.

The pump BHP, the power input, and the power losses from the power line to the pump can be calculated for each method of flow control. A summary of these calculations is shown in Table 7.4 for three flow conditions: full flow (100%), 75% flow, and 50% flow.

The annual power savings and present net worth of the savings for each system can be determined for a specific set of conditions. To continue the sample calculations, set the following conditions:

Annual operating hours, 8000 hr
Operating cycle:

full flow, 50% of the time
75% flow, 30% of the time
50% flow, 20% of the time

Full-flow BHP, 125 hp
Initial power rate, $0.06/kWh
Annual increase in power rates, 15%
Cost of money, 20%
Tax rate, 40%

TABLE 7.5 Annual Energy Savings and Net Present Worth for Adjustable-Speed Pumping Systems[a]

System	Annual savings, kWh	Net present worth, $
Fluid drive or eddy current system	120,056	30,907
AC variable-frequency system	179,536	45,448
Wound rotor with slip recovery system	183,104	47,141

[a] Base for comparison: throttling system.

The summary of these calculations in Table 7.5 shows the savings that can be achieved by means of adjustable-speed pumping systems compared to a throttling-type system. These calculations indicate that substantial power savings can be achieved with variable-speed types of pumping systems. The selection of the most effective system depends on the comparison of the system cost to the net present worth of the system savings and on application factors such as environment and maintenance.

7.5 APPLICATIONS TO CONSTANT-TORQUE LOADS

A wide variety of loads can be considered constant-torque loads. Conveyers are probably the most common type of constant-torque application. The means of achieving adjustable speed for these types of loads has been basically mechanical methods or DC drives. The adjustable-frequency induction motor drive systems add a new dimension to the method of adjusting the output speed for such loads.

A comparison of the annual power costs for the following adjustable-speed drive systems driving a constant-torque load is made in the following example:

Eddy current or hydrodynamic drives
Mechanical adjustable-speed drives
Adjustable-frequency induction motor drives

EXAMPLE

Consider a conveyer operated over approximately a 3:1 speed range with a constant-torque requirement of 30 lb-ft.

Annual operating hours: 2000
Duty cycle:

1750 rpm, 1200 hr
1200 rpm, 500 hr
600 rpm, 300 hr

Power rates: $0.06/kWh

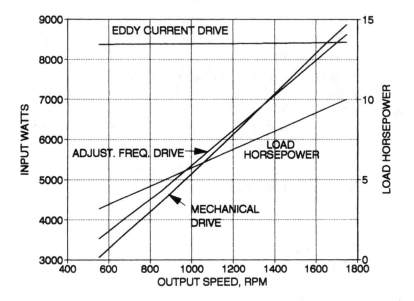

FIGURE 7.61 Input watts for a constant-torque load with various adjustable-speed drive systems.

TABLE 7.6 Annual Power Costs for Example Constant-Torque Conveyer Drive Systems

System	Annual energy consumption, kWh	Annual power cost, $
Eddy current drive	16,817	1,008.99
Mechanical adjustable-speed drive	14,649	878.94
Adjustable-frequency drive	14,544	872.64

Figure 7.61 shows the input power and load horsepower for each of the above drive systems over the speed range considered. The input power is based on a 10-hp, 1800-rpm energy-efficient drive motor for each system and the efficiency of each system.

Based on these data, the annual power consumption in kilowatt-hours and the power cost in dollars for each of the above systems are shown in Table 7.6. With the narrow range in annual power cost, additional factors such as initial cost, ease in operation, remote control features, environment, and maintenance must be considered in selecting the proper adjustable-speed drive system for a constant-torque load.

8

Brushless DC Motor Drives

In conventional DC motors with brushes, the field winding is on the stator and armature winding is on the rotor. Because of the brushes, the motor is expensive and needs maintenance. In addition, accumulation of the brush debris, dust, and commutator surface wear as well as arcing cannot be permitted in certain hazardous locations, which limits the application of DC brushed motors. As solid-state switching devices have been developed, it became possible to replace the mechanical switching components (commutator and brushes) by electronic switches. In fact, a brushless DC (BLDC) motor has a permanent magnet rotor and a wound field stator, which is connected to a power electronic switching circuit. Rotor position information is required for the power electronic driver. Figure 8.1 shows brushless DC motor drives and Fig. 8.2 a typical BLDC motor.

According to the National Electrical Manufacturers Association (NEMA), "a brushless DC motor is a rotating self-synchronous

FIGURE 8.1 Brushless DC motor drives. (Courtesy MPC Products Corporation, Skokie, IL.)

machine with a permanent magnet rotor and with known rotor shaft positions for electronic commutation." The advantage of brushless configuration in which the rotor (field) is inside the stator (armature) is simplicity of exiting the phase windings. Due to the absence of brushes, motor length is reduced as well. The disadvantages of the brushless configuration relative to the commutator motor are increased complexity in the electronic controller and need for shaft position sensing.

Main advantages of the BLDC motor drives are high efficiency, low maintenance and long life, low noise, control simplicity,

FIGURE 8.2 Typical BLDC motor, developed by Infranor Inc., Naugatuck, CT.

low weight, and compact construction. On the other hand, the main disadvantages of the BLDC motor drives are high cost of the permanent magnet materials, the problem of demagnetization, and limited extended speed, constant power range (compared to a switched reluctance machine).

Brushless DC motors can be classified based on the shape of their back EMF: trapezoidal or sinusoidal. In a BLDC motor with trapezoidal back EMF, the permanent magnets produce an air gap flux density distribution that is of trapezoidal shape. These motors, compared to motors with sinusoidal shape back EMF, have higher torque and larger torque ripples. They are also cheaper and used for general applications. In a BLDC motor with sinusoidal back EMF, the permanent magnets produce an air gap flux density distribution that is sinusoidal. These motors, compared to motors with trapezoidal shape back EMF, have smaller torque ripples. They are also expensive and are used for servo applications.

8.1 BLDC MACHINE CONFIGURATIONS

Permanent magnet (PM) BLDC machines can be classified according to the type of the permanent magnet (ferrite, ceramic, alnico, or rare earth), the shape of the back EMF waveform (trapezoidal or sinusoidal), the mounting of the permanent magnet (surface mounted or interior-mounted), the direction of the magnetic flux (radial or axial), the configuration of the windings (slot windings or slotless surface windings), or the power electronic drive (unipolar drive or bipolar drive).

Permanent magnet BLDC motors are classified as surface-mounted permanent magnet (SMPM) or interior-mounted permanent magnet (IPM) types. Figure 8.3 shows these two configurations. In the case of the surface-mounted permanent magnet machine, the magnets are mounted on the surface of the rotor, while for the interior-mounted permanent magnet machine, the magnets are inside the rotor. In the surface-mounted machine, the air gap might be nonuniform, while for the interior mounted machine, the air gap is uniform.

The stator shape of the PM BLDC motors can be configured with slots (slotted type) or without slots (slotless type); the slotless

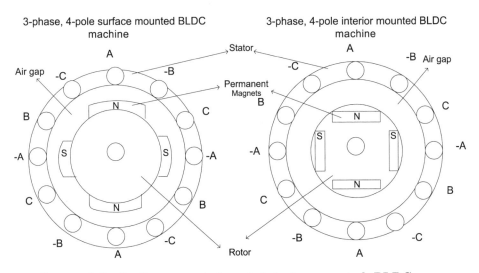

FIGURE 8.3 Surface-mounted and interior-mounted BLDC machines.

windings are also called surface winding or air gap winding. There is a special trend to use slotless BLDC motors especially in low-power and high-speed, high-performance applications. A method to eliminate the cogging torque in BLDC machines is to eliminate its source: the reluctance changes in the magnetic circuit during the rotation of the PM around the stator. In a traditional brushless motor, copper wires are wound through slots in a laminated steel core. As magnets pass by the lamination shoes, they have a greater attraction to the iron at the top of the laminations than to the air gap between shoes. This uneven magnetic pull causes cogging, which in turn increases motor vibrations and noise. Therefore, the key to smooth brushless performance centers on a slotless stator. Additionally, a slotless design significantly reduces damping losses.

Figure 8.4 shows the permanent magnet rotor of a three-phase, 1-hp BLDC motor; Fig. 8.5 shows the stator of a three-phase, 1-hp BLDC motor; and Fig. 8.6 shows the rotor and stator of a three-phase, 1-hp BLDC motor.

The BLDC motor consists of a stator, which contains the stator windings, and a rotor, on which permanent magnets are mounted. These magnets supply the field flux. The classic converter

FIGURE 8.4 Permanent magnet rotor of a three-phase, 1-hp BLDC motor.

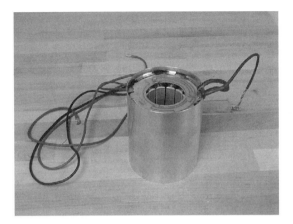

FIGURE 8.5 Stator of a three-phase, 1-hp BLDC motor.

for the BLDC motor is nothing but a DC/AC three-phase inverter, as shown in Fig. 8.7. The DC link voltage can be obtained from a pulse width modulation (PWM) rectifier from the AC supply, as shown in Fig. 8.8. A simple diode bridge can also be used.

To switch the coils in the correct sequence and at the correct time, the position of the rotor field magnets must be known. For

FIGURE 8.6 Rotor and stator of a three-phase, 1-hp BLDC motor.

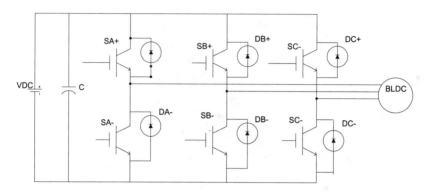

FIGURE 8.7 Classic BLDC power electronic driver.

locating the rotor field magnets, an absolute sensing system is required. An absolute sensing system may consist of Hall sensors or optical encoders.

The function of the controller is to switch the right currents in the right stator coils at the right time in the right sequence by taking the information supplied by the sensor and processing it with preprogrammed commands to make the motor perform as desired.

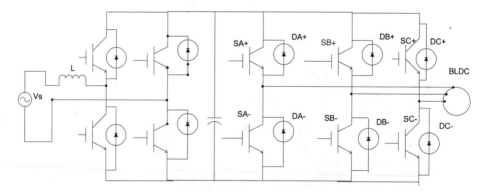

FIGURE 8.8 A typical BLDC driver with a PWM rectifier at its front end.

8.2 MODELING

A BLDC machine model is shown in Fig. 8.9. As shown in this figure, R is the phase resistance; L is the phase inductance; and EA, EB, and EC are phase back EMF voltages. The waveforms are shown in Fig. 8.10.

Motor terminal voltage for a three-phase full bridge inverter with six switches and Y-connected motor can be expressed as follows:

$$V_{DC} = E_f + 2\,R\,i_a + 2\,L\frac{di_a}{dt}$$

Assuming switches are ideal and the EMF between conducting phases is constant (trapezoidal EMF), the instantaneous armature current can be written as follows:

$$i_a(t) = \frac{V_{DC} - E_F}{2\,R}\left(1 - e^{\frac{R}{L}t}\right) + I_0 e^{\frac{R}{L}t}$$

We can express the EMF simply as a function of rotor speed:

$$E_f = k_e\,n$$

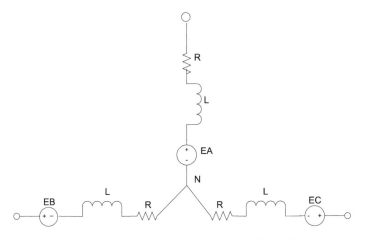

FIGURE 8.9 Equivalent circuit of a BLDC machine.

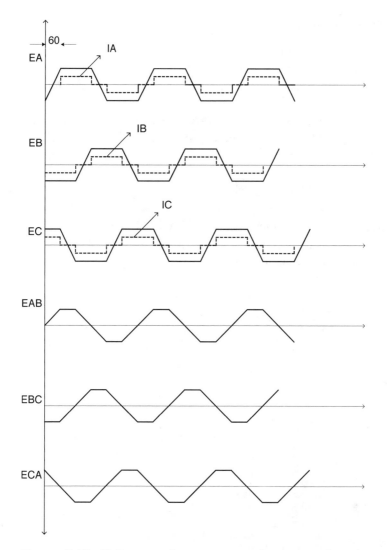

FIGURE 8.10 Voltage and current waveforms as a function of rotor position.

Torque equation is similar to a conventional DC commutator motor and can be written as follows:

$$T_d = k_t \, i_a$$

The average developed torque can be maximized and torque ripple can be minimized if the EMF voltage waveform has a trapezoidal shape. Trapezoidal waveform can be achieved by switching the MOSFETs or IGBTs in such a way that two-phase windings are always connected in series during the whole conduction period of 60 degrees. In addition, proper shaping and magnetizing of the permanent magnets and stator windings are important factors to obtain the trapezoidal waveform. Owing to manufacturing tolerances, armature reactions, and other parasitic effects, the EMF waveform is never ideally flat. However, a torque ripple below 5% can be achieved.

Power density and torque density are the measures to judge how the active materials of a BLDC motor are best utilized. NdFeB

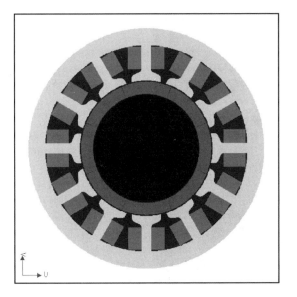

FIGURE 8.11 Geometry of a three-phase, 1-hp BLDC motor.

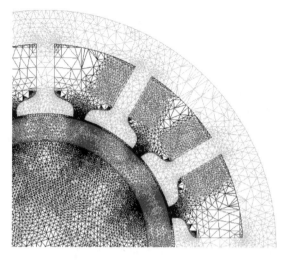

FIGURE 8.12 Enlarged view of the electromagnetic mesh of the three-phase, 1-hp BLDC motor.

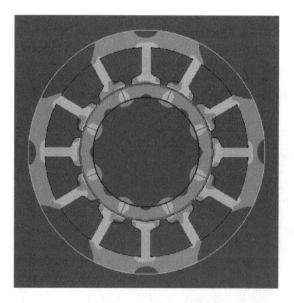

FIGURE 8.13 Magnetic flux density of the three-phase, 1-hp BLDC motor.

magnets offer the highest energy density at reasonable costs. Their major drawback, compared to SmCo, is temperature sensitivity. When using NdFeB magnets, the motor's temperature must be kept below 170–250°. Since the rotor losses are small, passive cooling is employed for the rotor. The main dimensions (inner stator diameter and effective length of core) for the BLDC are determined by rated power output, air gap magnetic flux density, and armature line current. Geometry of a three-phase, 1-hp BLDC motor is shown in Fig. 8.11.

Using a finite element (FE) software package such as Maxwell 2D (Ansoft Corp.) a BLDC machine can be analyzed and various solutions presented. The software generates an initial mesh and then refines the solution to achieve the required precision. An enlarged view of the final mesh is shown in Fig. 8.12. Magnetic flux density is shown in Fig. 8.13.

8.3 BLDC POWER ELECTRONIC DRIVERS

The permanent magnet BLDC motor shares the same torque-speed characteristics and the basic operating principles of the brushed DC

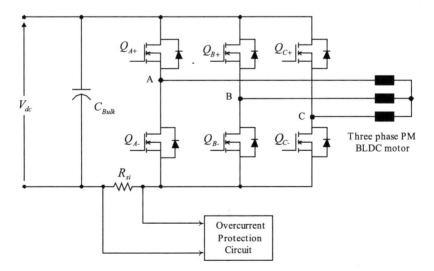

FIGURE 8.14 Three-phase MOSFET inverter.

motor. The main difference is that the field windings are replaced by permanent magnets and the commutation is done electronically. Using permanent magnets and eliminating the brushes offers many distinctive advantages, such as high performance torque control, low torque ripples, long life, high power-to-weight ratio, low noise and low electromagnetic interference, better heat dissipation, low maintenance, and very high speed of operation.

The power electronic driver for the BLDC motor can be an IGBT-based inverter, as shown in Figs. 8.7 and 8.8. It can also be a MOSFET-based inverter as shown in Fig. 8.14. An IGBT-based driver is usually used for high-power or high-voltage applications.

Figure 8.15 shows the three-phase voltage and current waveforms for a trapezoidal back EMF BLDC machine. In order to maintain the current as shown in Fig. 8.15, a hysteresis (bang-bang) control technique can be used. With the hysteresis control, the

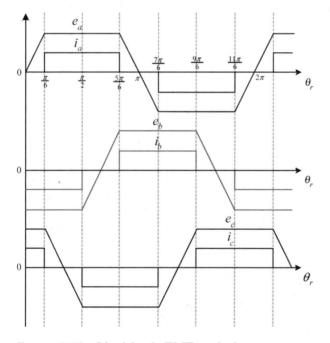

FIGURE 8.15 Ideal back EMF and phase current waveforms of the BLDC motor.

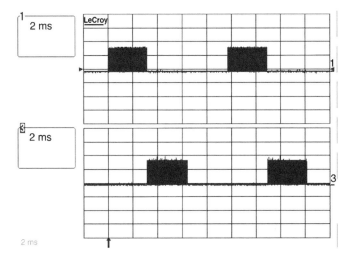

FIGURE 8.16 Driving pulses for Q_{A-} and Q_{B-} of inverter in Fig. 8.14, 10 V/div.

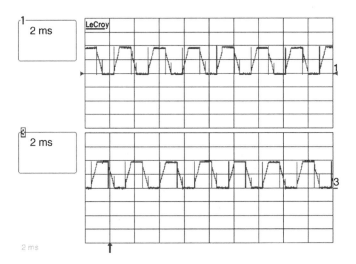

FIGURE 8.17 Voltages of phase A and phase B of the driver shown in Fig. 8.14.

current is directly controlled; therefore, output torque of the motor is controlled. However, the switching frequency is variable. In order to have a constant switching frequency, a pulse width modulation technique can be used. There are many integrated circuits (ICs) available in the market that use PWM techniques. The advantage is that the average of the applied voltage to the motor terminals is directly controlled. Therefore, speed of the motor is directly controlled by changing the duty cycle of the PWM switching scheme. Simplicity is another advantage of the PWM methods.

In Fig. 8.14, for overcurrent protection, sense resistance R_{si} is used. The sense voltage over R_{si}, which is proportional to the load current, is fed into a comparator on the main control board. The overcurrent circuit limits the motor current. Driving pulses and phase voltages for the inverter in Fig. 8.14 are given in Figs. 8.16 and 8.17, respectively.

8.4 SENSORLESS TECHNIQUES FOR BLDC MOTOR DRIVES

Brushless DC motor drives require rotor position information for proper operation. Position sensors are usually used to provide the position information for the driver. However, in sensorless drives, position sensors are not used. Instead, position information is obtained indirectly. Advantages of sensorless drives include increased system reliability, reduced hardware cost, reduced feedback units, and decreased system size. In addition, they are free from mechanical and environmental constraints. However, sensorless techniques may affect system performance. Low-speed sensorless operation is also difficult.

Many model-based sensorless techniques have been proposed. In these methods, a model of the machine is used to obtain the position information from measured signals such as voltages and currents. Based on the model, usuallya linear or nonlinear equation for position is solved. For example, in a d-q model–based sensorless technique, the actual d-q transformed currents and voltages, those on a hypothetical axis offset from the d-q axis by a small angle $\Delta\theta$, the output voltages of the model on the hypothetical axis, and those on the actual d-q axis are compared. The difference between the calculated voltages of the hypothetical axis considering $\Delta\theta = 0$ and the actual d-q axis voltages on the hypothetical axis considering $\Delta\theta$ gives the actual change in rotor position from the previously known position.

Many sensorless techniques are based on the back EMF of the machine. Generally, these techniques are used for BLDC machines with trapezoidal back EMF. For example, the zero crossing point of the back EMF voltage can be detected. In this scheme, the three terminal voltages and neutral voltage of the motor with respect to the negative DC bus voltage are measured. The terminal voltage is equal to the neutral voltage at the instants of the zero crossing of the back EMF waveform. In order to use the zero crossing point to derive the switching sequence, this point has to be shifted by 30 degrees. Therefore, in this method, speed estimation is required.

In the third-harmonic back EMF sensing technique, the position of the rotor can be determined based on the stator third-harmonic voltage component. To detect the third-harmonic voltage, a three-phase set of resistors is connected across the motor windings (Fig. 8.18). The voltage across the points P and Q is denoted by E3, and it determines the third-harmonic voltage. This voltage is integrated for a zero crossing detector. The output of the zero crossing detector determines the switching sequence for turning on the switches. The resistances and inductances are shown in the circuit of Fig. 8.18. The important point is that the summed terminal voltages contains only the third and the multiples of the third harmonic due to the fact that only zero sequence current components can flow through the motor neutral. This voltage is dominated by the third harmonic.

By integrating the voltage E3, we get the third-harmonic flux linkage λ_3. The third-harmonic flux linkage lags the third harmonic of the phase back EMF voltage by 30 degrees. The zero crossings of the third harmonic of the flux linkage correspond to the commutation instants of the BLDC driver, as shown in Fig. 8.19.

The back EMF integration method is another sensorless technique. In this method, by integrating the back EMF of the unexcited phase, position information is obtained. The integration of the back EMF starts when the back EMF of the open phase

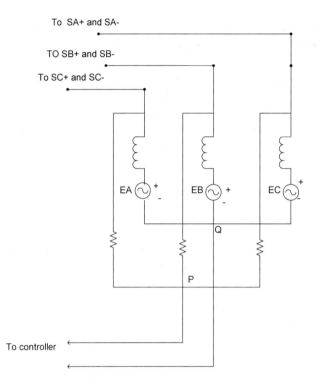

FIGURE 8.18 Resistors for the third harmonic back EMF sensing technique.

crosses zero. Here, speed estimation is not required. A threshold is set to stop the integration, which corresponds to a commutation point. Assuming trapezoidal back EMF, the threshold voltage is kept constant throughout the speed range. Integrating from the zero crossing point (ZCP) to the commutation point (CP) is constant (not a function of speed). Therefore, speed estimation is not required.

The freewheeling diode conduction-sensing technique uses indirect sensing of the phase back EMF to obtain the switching instants of the BLDC motor. Considering the 120-degree conducting wye-connected BLDC motor, one of the phases is always open. After opening the phase for a short interval of time, there remains a phase current flowing through a freewheeling diode. This open

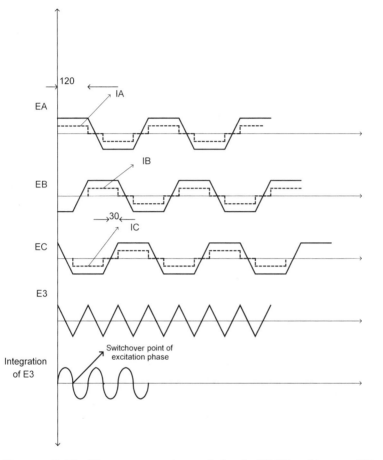

FIGURE 8.19 Phase currents and back EMF voltages, E3, and integration of E3 versus rotor position.

phase current becomes zero in the middle of the commutation interval, which corresponds to the point where the back EMF of the open phase crosses zero.

Some sensorless techniques are also based on the magnetic flux. These methods are usually used for BLDC motors with sinusoidal back EMF. For example, in flux integration methods, instantaneous flux is obtained from the integration of the voltage equation of the machine. By knowing the initial position, the relationship of the flux linkage to the rotor position, and machine parameters, the rotor position is estimated. The speed is determined by the rate of change of the flux linkage from the integration results.

Observer-based sensorless methods are generally used for BLDC motors with sinusoidal back EMF. In these techniques, a Kalman filter might be used. The Kalman filter provides an optimum observation from noisy sensed signals and processes that are disturbed by random noise. A mathematical model describing the motor dynamics is known. The rotor position can be determined based on the voltages and currents. The measured voltages and currents are transformed to stationary frame components. Using the state equations and a Kalman filter, the missing states (rotor position and velocity) are estimated. The estimated rotor position is used for commutation. The function of the filter is to correct the estimation process in a recursive manner. The filter constantly works on the output and corrects its quality based on the measured values. Based on the deviation from the estimated value, the filter provides an optimum output value at the next output instant.

In a state observer, the output is defined as a combination of the states. This output is compared with the equivalent measured output of the real motor. Any error between the two signals is used to correct the state trajectory of the observer. The accuracy of the position information depends on the stability of the observer. For the system to be stable, the gain of the system has to be optimized. Furthermore, initial information of the states is required for proper convergence of the observer.

SELECTED READINGS

1. Miller, T. J. E. (1993). *Brushless Permanent Magnet and Reluctance Motor Drives*. Madison, WI: Magna Physics Publishing.
2. Hendershot, J. R., Miller, T. J. E. (1994). *Design of Brushless Permanent-Magnet Motors*. Oxford, UK: Oxford.
3. Krishnan, R. (2001). *Electric Motor Drives: Modeling, Analysis, and Control*. Upper Saddle River, NJ: Prentice-Hall.
4. Johnson, J. P. (1998). Synchronous-misalignment detection/cor-

rection technique of sensorless BLDC control. Ph.D. dissertation, Texas A&M University.

5. Skvarenina, T. L. (2002). *The Power Electronics Handbook*. Boca Raton, FL: CRC Press.

6. Matsui, N. (April 1996). Sensorless PM brushless DC motor drives. *IEEE Trans. on Industrial Electronics* 43:300–308.

7. Matsui, N. (1993). Sensorless operation of brushless DC motor drives. In: Proc. IEEE Conf. on Industrial Electronics, Control, and Instrumentation. Vol. 2. pp. 739–744.

8. Senjyu, T., Uezato, K. (1995). Adjustable speed control of brushless DC motors without position and speed sensors. In: Proc. IEEE/IAS Conf. on Industrial Automation and Control: Emerging Technologies. pp. 160–164.

9. Sicot, L., Siala, S., Debusschere, K., Bergmann, C. (1996). Brushless DC motor control without mechanical sensors. IEEE Power Electronics Specialist Conf. pp. 375–381.

10. Iizuka, K., Uzuhashi, H., Kano, M. (May/June 1985). Microcomputer control for sensorless brushless motors. *IEEE Trans. on Industry Applications* IA-27:595–601.

11. Consoli, A., Musumeci, S., Raciti, A., Testa, A. (Feb. 1994). Sensorless vector and speed control of brushless motor drives. *IEEE Trans. on Industrial Electronics* 41:91–96.

12. Wu, R., Slemon, G. R. (Sep./Oct.1991). A permanent magnet motor drive without a shaft sensor. *IEEE Trans. on Industry Applications* 27:1005–1011.

13. Ertugrul, N., Acarnley, P. (Jan./Feb. 1994). A new algorithm for sensorless operation of permanent magnet motors. *IEEE Trans. on Industry Applications* 30:126–133.

14. Takeshita, T., Matsui, N. (1994). Sensorless brushless DC motor drive with EMF constant identifier. In: Proc. IEEE Conf. on Industrial Electronics, Control, and Instrumentation. Vol. 1. pp. 14–19.

15. Matsui, N., Shigyo, M. (Jan./Feb. 1992). Brushless DC motor control without position and speed sensors. *IEEE Trans. on Industry Applications* 28:120–127.

16. Watanabe, H., Katsushima, H., Fujii, T. (1991). An improved measuring system of rotor position angles of the sensorless direct drive servomotor. In: Proc. IEEE 1991 Conf. on Industrial Electronics, Control, and Instrumentation. pp. 165–170.

17. Kim, J. S., Sul, S. K. (1996). New approach for the low speed

operation of the PMSM drives without rotational position sensors. *IEEE Trans. on Power Electronics* 11:512–519.

18. Oyama, J., Abe, T., Higuchi, T., Yamada, E., Shibahara, K. (1995). Sensorless control of a half-wave rectified brushless synchronous motor. In: Conf. Record of the 1995 IEEE Industry Applications Conf. Vol. 1. pp. 69–74.

19. Wijenayake, A. H., Bailey, J. M., Naidu, M. (1995). A DSP-based position sensor elimination method with on-line parameter identification scheme for permanent magnet synchronous motor drives. In: IEEE Conf. Record of the 13th Annual IAS Meeting. Vol. 1. pp. 207–215.

20. Kim, J. S., Sul, S. K. (1995). New approach for high performance PMSM drives without rotational position sensors. In *IEEE Conf. Proc. 1995, Applied Power Electronics Conf. and Exposition.* Vol. 1. pp. 381–386.

21. Schrodl, M. (1994). Sensorless control of permanent magnet synchronous motors. *Electric Machines and Power Systems* 22: 173–185.

22. Brunsbach, B. J., Henneberger, G., Klepsch, T. (1993). Position controlled permanent magnet excited synchronous motor without mechanical sensors. In: IEEE Conf. on Power Electronics and Applications. Vol. 6. pp. 38–43.

23. Dhaouadi, R., Mohan, N., Norum, L. (July 1991). Design and implementation of an extended Kalman filter for the state estimation of a permanent magnet synchronous motor. *IEEE Trans. on Power Electronics* 6:491–497.

24. Sepe, R. B., Lang, J. H. (Nov./Dec. 1992). Real-time observer-based (adaptive) control of a permanent-magnet synchronous motor without mechanical sensors. *IEEE Trans. on Industry Applications* 28:1345–1352.

25. Senjyu, T., Tomita, M., Doki, S., Okuma, S. (1995). Sensorless vector control of brushless DC motors using disturbance observer. In: PESC '95 Record, 26th Annual IEEE Power Electronics Specialists Conf. Vol. 2. pp. 772–777.

26. Solsona, J., Valla, M. I., Muravchik, C. (Aug. 1996). A nonlinear reduced order observer for permanent magnet synchronous motors. *IEEE Trans. on Industrial Electronics* 43:38–43.

27. Kim, Y., Ahn, J., You, W., Cho, K. (1996). A speed sensorless vector control for brushless DC motor using binary observer. In: Proc. of the 1996 IEEE IECON 22nd Int'l. Conf. on Industrial

Electronics, Control, and Instrumentation. Vol. 3. pp. 1746–1751.

28. Furuhashi, T., Sangwongwanich, S., Okuma, S. (April 1992). A position-and-velocity sensorless control for brushless DC motors using an adaptive sliding mode observer. *IEEE Trans. on Industrial Electronics* 39:89–95.

29. Hu, J., Zhu, D., Li, Y., Gao, J. (1994). Application of sliding observer to sensorless permanent magnet synchronous motor drive system. In: IEEE Power Electronics Specialist Conf. Vol. 1. pp. 532–536.

30. Moreira, J. (Nov./Dec. 1996). Indirect sensing for rotor flux position of permanent magnet AC motors operating in a wide speed range. *IEEE Trans. on Industry Applications Society* 32:401–407.

31. Ogasawara, S., Akagi, H. (Sep./Oct. 1991). An approach to position sensorless drive for brushless DC motors. *IEEE Trans. on Industry Applications* 27:928–933.

32. Jahns, T. M., Becerra, R. C., Ehsani, M. (Jan. 1991). Integrated current regulation for a brushless ECM drive. *IEEE Trans. on Power Electronics* 6:118–126.

33. Becerra, R. C., Jahns, T. M., Ehsani, M. (March 1991). Four-quadrant sensorless brushless ECM drive. *IEEE Applied Power Electronics Conf. and Exposition.* 202–209.

9

Switched Reluctance Motor Drives

Switched reluctance machines (SRMs) have salient stator and rotor poles with concentrated windings on the stator and no winding on the rotor. The stator windings can be wound externally and then slid onto the stator poles. This provides for a very simple manufacturing process, thus the cost of the machine is also low. The SRM has a single rotor construction, essentially made of stacks of iron, and does not carry any coils or magnets. This feature gives it a rugged structure and provides the machine with the advantage that it can be used at high speeds and better withstand high temperatures. The SRM achieves high torque levels at low peak currents by using small air gaps. Figures 9.1 and 9.2 show a cross-sectional views of a typical 6/4 SRM with six stator and four rotor poles.

The choice of the number of poles to be used in the SRM is important due to the vibration that is produced. A structure such as 12/8 (12 stator and eight rotor poles) provides lower mechanical vibration compared to the 6/4 structure. SRM machines are well suited for high-speed operations and tend to have higher efficiency

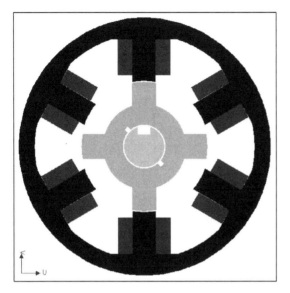

FIGURE 9.1 Cross section of a typical 6/4 SRM with six stator and four rotor poles.

at high speeds. They provide constant power over a wide speed range and are highly dynamic with speed. In fact, switched reluctance motor drives are inherently adjustable- or variable-speed drives.

The SRM is a highly reliable machine as it can function even under faulty conditions with reduced performance. One of the reasons for this is that the rotor does not have any excitation source and thus does not generate power into the faulted phase; therefore, no drag torque would be produced under the motoring mode and there are no sparking/fire hazards due to excessive fault currents. In addition, the machine windings are both physically and electromagnetically isolated from one another, reducing the possibility of phase-to-phase faults. The classic SRM drive as a system with converter involves two switches and a winding in series. Thus, even in a case of both switches being turned on at the same time, no shoot-

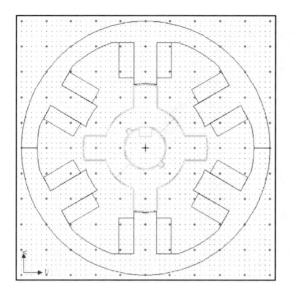

FIGURE 9.2 Cross sectional view of a typical 6/4 SRM with six stator and four rotor poles.

through faults would occur, unlike the case of AC drives, which lead to shorting of the DC bus.

Switched reluctance machines by a nonsinusoidal voltage waveform, thus resulting in a high torque ripple. This also leads to high noise levels. As per the standards, to have torque ripples of less than 15% in nonsinusoidal excited machines, the number of phases in the SRM has to be increased, which would reflect on the system cost due to the increased number of parts, in this case the switches. A better solution is to use advanced control techniques to reduce the torque ripple and noise. The SRM converters have very high efficiency at low and high speeds. During generation, the efficiency values are remarkably high—above 90%—over a wide speed range.

Electromagnetic torque in the SRM is produced by the tendency of the salient rotor poles to align with the excited stator poles and attain the least reluctance position. The torque developed depends on the relative position of the phase current with respect to the inductance profile. If the current falls on the negative slope of the inductance profile, then the machine is in the generating mode. The back EMF developed depends on the magnetic parameters of the machine, rotor position, and the geometry of the SRM.

The current waveforms for the motoring mode and the generating modes are mirror images of each other. During generation, initial excitation has to be provided from an external source, it being a single excited structure. The stator coils are turned on around the aligned position and then turned off before unalignment for generating electricity. The turn-on and turn-off angles as well as the current determine the performance of the machine. During the high-speed motoring mode, the peak value of the current depends on the turn-on time. These timings greatly help in designing optimal and protective control. In addition, in SRM, by changing the turn-on and turn-off angles, the system may be optimized to operate under maximum efficiency, minimum torque ripple, or minimum ripple DC link current. The extended constant power/speed ratio capability of the SRM enables less power requirement during the motoring mode.

The SRM can be current controlled for both motoring and generating modes of operation. During motoring the current is controlled by adjusting the firing angles and applying the current during the magnetization period. If during the current control there is overlapping of the phase currents, it leads to an increase in the maximum torque level. During the generation mode of operation, the torque must be fed to the machine when the inductance level is reducing, i.e., when the rotor is moving from the aligned position to the unaligned position.

9.1 HISTORY OF SWITCHED RELUCTANCE MACHINE

The guiding principle behind the SRM drive is that when a magnetically salient rotor is subject to the flow of flux in the magnetic circuit, it tends to move toward the position of minimum reluctance. This phenomenon has been known ever since the first experiments on electromagnetism. In the first half of the 19th century, scientists all over the world were experimenting with this effect to produce continuous electrical motion. A breakthrough came from W. H. Taylor in 1838, who obtained a patent for an electromagnetic engine in the United States. This machine was composed of a wooden wheel on the surface, on which was mounted seven pieces of soft iron equally spaced around the periphery. The wheel rotated freely in the framework in which four electromagnets were mounted. These magnets were connected to a battery through a mechanical switching arrangement on the shaft of the wheel such that excitation of an electromagnet would attract the nearest piece of soft iron, turning the wheel and energizing the next electromagnet in the sequence to continue the motion. However, the torque pulsations were the main drawback of this machine compared to DC and AC machines.

Improvement was noticed with the introduction of electronic parts that replaced the mechanical arrangements. The trend moved toward reducing the mechanical arrangements and parts while increasing the electronic parts. Improved magnetic material and advances in machine design have brought the SRM into the variable-speed drive market. Presently demand for SRMs is increasing as they offer superior performance with lower price. Other than simplicity and low-cost machine manufacturing, the main motivation toward SRM use is the availability of low-cost power electronic switches, control electronic components, integrated circuits (ICs), microcontrollers, microprocessors, and digital signal processors (DSPs).

In the present global scenario, SRM drives are one of the major emerging technologies in the field of adjustable-speed drives. They have many advantages in terms of machine efficiency, power density, torque density, weight, volume, robustness, and operational flexibility. SRM drives are finding applications in general-purpose industrial drives, traction, devices domestic appliances, and office and business equipment. The emerging markets in consumer and industrial products are very cost sensitive as well as demand high reliability and performance, equivalent to DC and induction motor drives.

9.2 FUNDAMENTALS OF OPERATION

Switched reluctance machines are similar to synchronous machines but with significant distinctive features. They are referred to as doubly salient pole machines as their stator and rotor both have salient poles. This configuration proves to be more effective as far as electromagnetic energy conversion is concerned. Another prominent feature is that there is no coil or permanent magnet (PM) on the rotor. Figure 9.3 shows the rotor and stator of a 10-hp 6/4 SRM with six stator and four rotor poles.

The absence of permanent magnets or coils on the rotor means that torque is produced purely by the saliency of the rotor laminations. The torque is produced with respect to the direction of the flux through the rotor, and hence the direction of the flow of the current in the stator phase windings is not important. The need for unipolar phase current in the reluctance motor results in simpler and more reliable power converter circuits.

9.2.1 Torque Equation

Despite the simple operation of SRMs, an accurate analysis of the motor's behavior requires a relatively complex mathematical approach. The instantaneous voltage across the terminals of a single

FIGURE 9.3 Rotor and stator of a 10-hp SRM with six stator and four rotor poles.

phase of the motor winding is related to the flux linked in the winding by Faraday's law,

$$v = Ri + \frac{d\phi}{dt}$$

where v is the terminal voltage, i is the phase current, R is the motor resistance, and $d\phi/dt$ is the flux linked by the winding. We can rewrite this equation as follows:

$$v = Ri + \frac{d\phi}{di}\frac{di}{dt} + \frac{d\phi}{d\theta}\frac{d\theta}{dt}$$

where $d\phi/di$ is defined as the instantaneous inductance $L(\theta,i)$ and $d\phi/di$ is defined as the instantaneous back EMF $K_b(\theta,i)$. Figure 9.4 depicts the flux-current characteristic of the machine. The above equation defines the transfer of electrical energy to the magnetic field of the machine. The following equations describe the conversion of the field's energy into mechanical energy. Multiplying each side of Faraday's equation by the electrical current gives an expression for the instantaneous power:

$$vi = Ri^2 + i\frac{d\phi}{dt}$$

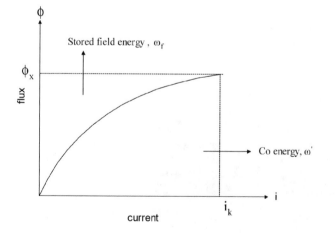

FIGURE 9.4 Flux-current characteristic of an SRM.

The left side of this equation represents the instantaneous electrical-power delivered to the SRM. The first term on the right side represents the resistive losses in the SRM stator winding. If power is to be conserved, then the second term on the right side must represent the sum of the mechanical power output of the machine and any power stored in the magnetic field.

$$i\frac{d\phi}{dt} = \frac{dW_m}{dt} + \frac{dW_f}{dt}$$

$$\frac{dW_m}{dt} = T\frac{d\theta}{dt}$$

Substituting the above torque term, we obtain

$$i\frac{d\phi}{dt} = T\frac{d\theta}{dt} + \frac{dW_f}{dt}$$

By solving the equations, we get

$$T(\theta, \phi) = i(\theta, \phi)\frac{d\phi}{d\theta} - \frac{dW_f(\theta, \phi)}{d\theta}$$

We can simplify to

$$T = -\frac{dW_f}{d\theta}$$

It is better to express torque in terms of the current rather then flux; therefore, torque is expressed in terms of coenergy instead of energy:

$$W_f = \int_0^\phi i(\theta, \phi)d\phi$$

$$W_c = \int_0^i \phi(\theta, i)di$$

W_c is defined as the magnetic field coenergy.

$$W_c + W_f = i\phi$$

Differentiation and substituting in the main torque equation,

$$dW_c + dW_f = \phi di + id\phi$$

$$T = \frac{id\phi - (\phi di + id\phi - dW_c(\theta, i))}{d\theta}$$

For simplicity, coenergy can be written as

$$dW_c(\theta, i) = \frac{dW_c}{d\theta} d\theta + \frac{dW_c}{di} di$$

$$T = \frac{dW_c}{d\theta}$$

After neglecting the saturation, we can write flux and current as

$$\phi = L(\theta)i$$

Substituting in the coenergy equation,

$$W_c = \frac{1}{2} i^2 L(\theta)$$

The simplified torque equation is obtained as follows:

$$T = \frac{1}{2} i^2 \frac{dL}{d\theta}$$

9.2.2 Characteristics of SRMs

The torque-speed characteristic of an SRM depends on the control used in the machine. It is also easily programmable, which makes the SRM attractive. However, there are some limitations due to the physical constraints such as the supply voltage and allowable temperature rise of the motor under increasing load.

As in other motors, torque is limited by the maximum allowed current and speed by the bus voltage. With increasing shaft speed, a current limit region persists until the rotor reaches a speed where the back EMF of the motor is such that, given the DC bus voltage limitation, we can get no more current in the winding, thus no more torque from the motor. At this point, which is called the base speed, we can obtain constant torque operation in which the shaft output power remains constant at its maximum. At still higher speeds, the back EMF increases and the shaft output power begins to drop. The product of the torque and the square of the speed remains constant in this region. The SRM holds an outstanding benefit over other machines as the ratio between the maximum speed and base speed in this machine is up to 10, which specifies that the SRM has extended speed constant power operation. The torque-speed characteristic is shown in Figure 9.5.

In SRM drives, the phase flux is the sum of the self-flux and the mutual flux. The self-flux is dominated as it goes through mostly iron and two air gaps across the phase poles, while the mutual flux goes to the air gap between the stator poles and interpolar part of the rotor poles. Due to the more effective air gap, there is high reluctance and lower contribution to the phase flux. Hence, the rotor position and related phase currents determine the self-inductance of the phase. Variations in the mutual inductance due to the phase current are practically negligible. There is no variation in self-inductance when the rotor is unaligned, but it decreases when the rotor is near the aligned position. The variation in the mutual inductance is not significant compared to the variations in self-inductance. The inductance characteristic with respect to different rotor positions

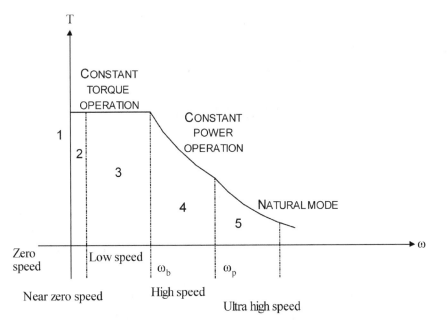

FIGURE 9.5 Torque-speed characteristic of an SRM.

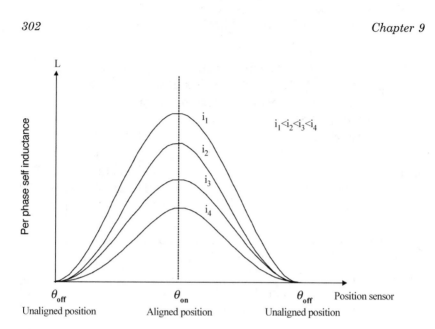

FIGURE 9.6   Self-inductance–rotor position characteristic of an SRM.

and different phase current levels is shown in Figure 9.6. Figure 9.7 depicts the flux-current characteristic of an SRM as a function of rotor position. Figures 9.8 – 9.10 show a flux density map, flux lines, and the wire mesh of a typical 6/4 SRM, respectively.

9.3 MACHINE CONFIGURATIONS

There are several factors that are important in selecting the number of phases and number of poles. The number of phases is determined by the following factors:

- *Starting capability.* When the rotor and stator poles are aligned, the single-phase machine cannot start; therefore, it requires a permanent magnet on the stator at an intermediate position to the stator poles to keep the rotor poles at the unaligned position.
- *Directional capability.* Two-phase machines are capable of only one direction of rotation, whereas three-phase machines are capable of rotation in both directions.

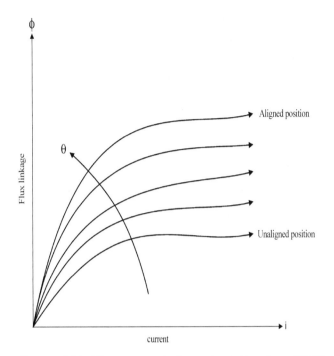

FIGURE 9.7 Flux-current characteristic of an SRM as a function of rotor position.

- *Reliability*. A higher number of phases gives higher reliability because a failure of one or more phases will allow the running of the machine with the remaining phases.
- *Cost*. A higher number of phases requires a higher number of converter phase units, their drivers, logic power supplies, and control units, which increases the overall cost.
- *Power density*. A higher number of phases provides higher power density.
- *Efficient high-speed operation*. Efficiency can be increased by reducing the core losses at high speed by decreasing the number of stator phases and lowering the number of phase switching per revolution.

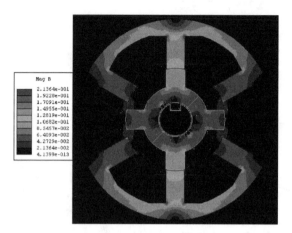

FIGURE 9.8   Flux density map of a typical 6/4 SRM.

In order to select the number of poles for the rotor and stator, the ratio between stator and rotor poles should be a noninteger. Different combinations of stator and rotor poles are given in Table 9.1 for typical SRM drives.

The limiting factors in pole selection are the number of power converter switches and their associated cost of gate drives and logic

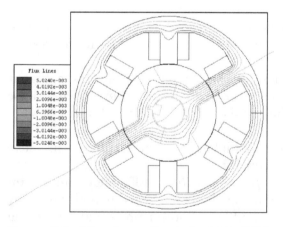

FIGURE 9.9   Flux lines of a typical 6/4 SRM.

FIGURE 9.10 Wire mesh of a typical 6/4 SRM.

supplies. With the increase in the number of poles, the cost increases due to increased winding insertion cost and terminal cost. Stator frequency is proportional to the number of poles; hence, with the increase in the number of poles, stator frequency increases, which results in higher core losses and considerably greater conduction time to provide rise and fall of the current compared to that of a machine with fewer poles. If the number of rotor poles is less, the copper losses increase and so does the phase conduction overlap. Due to the increased switching frequency, the commutation torque ripple frequency is also increased, thus making its filtering easier. Overlapping phase conductions and their effective control can

TABLE 9.1 Typical Numbers of Stator and Rotor Poles

Stator	6	12	8	12	24
Rotor	4	8	6	10	16

attenuate the commutation torque ripple magnitude, leading to quiet operation. Figures 9.11 and 9.12 show cross-sectional views of a typical 10/6 SRM and a typical 6/4 SRM, respectively.

Switch reluctance machines can offer a wide variety of aspect ratios and salient pole topologies. Single-phase motors are the simplest SRMs with the fewest connections between machine and electronics. The disadvantages lie in very high torque ripple and inability to start at all angular positions. This configuration is attractive for very high-speed applications, but starting problems may preclude their use.

For two-phase motors, the problems of starting compared with the single-phase machines can be overcome by stepping the air gap or providing asymmetry in the rotor poles. This machine may be of interest where the cost of winding connections is important, but again high torque ripple is the disadvantage.

Three-phase motors offer the simplest solution to starting and torque ripple without resorting to high numbers of phases. Hence, 6/4 SRM has been the most popular topology. Alternative three-phase machines with doubled pole numbers can offer a better solution for lower-speed applications. However, torque ripple, especially in the voltage control single-pulse operating mode, is again a drawback.

Four-phase motors are popular for reducing torque ripple further; however, the large number of power devices and connections renders four-phase motors to limited applications. Five-phase

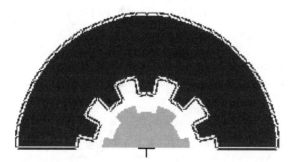

FIGURE 9.11 Half cross section of a typical 10/6 SRM with 10 stator and six rotor poles.

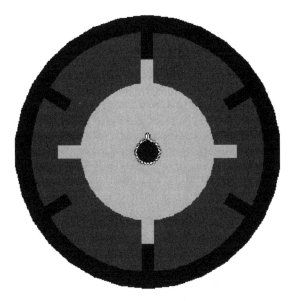

FIGURE 9.12 Cross section of a 6/4 SRM.

and six-phase motors can offer better torque ripple reduction compared with the four-phase and three-phase SRMs.

9.4 DYNAMIC MODELING OF SRMs

Recently, SRMs have been receiving much attention due to their various advantages. First, it has no windings or permanent magnets on the rotor. Thus, it can be mass produced at a fairly low cost compared to the other motor types. Furthermore, the SRM has additional advantages of low inertia, minimal losses on the rotor, and mechanical robustness, which allow it to be driven at high speeds. In addition, the stator windings are concentrated, making it easier to wind compared to the other AC and DC machines. Hence, SRM drive systems have been found to be suitable for home appliances, industrial applications, automotive applications, aircraft starter/generator systems, washing machines, compressors, electric/hybrid electric vehicles, etc.

However, despite its simple appearance, it is more difficult to design due to its nonlinearity. Due to these nonlinearities present in the system, analyzing the complete SRM drive becomes a cumbersome task. Furthermore, simulation of the SRM drive system requires proper modeling of the SRM subsystems. In this regard, various methods of estimating the torque and current of SRM drives have been introduced. In previous research carried out in this area, detailed nonlinear models of variable reluctance motors have been developed. The nonlinear characteristics of flux linkage–current–position are represented by corresponding analytical expressions. In order to obtain such exhaustive results, either through experimental verification or by finite element analysis (FEA) methods, a huge amount of time is required.

In addition, state space models for SRM drives have been developed wherein the phase currents are specified as the state variables (whereas the inductances are specified as parameters). Basically, this method is a more accurate way of representing the phase inductances in the model. Other research work in this area involves simulation of electromechanical energy conversion and the drive control system. Again, as is the case with other modeling methods, this too needs complete information of the inductance–current–position characteristics. Furthermore, this method of modeling does not consider low-speed operation.

Prediction of the dynamic performance of the SRM in various operating regions with accurate results requires a fairly accurate representation of its magnetic characteristics. This means the variation of flux linkage with rotor position and current must be accurately determined. Basically, this section aims at presenting a comprehensive dynamic model for the SRM drive system, which covers both low-speed and high-speed modes of operation. Such a detailed dynamic model takes care of the magnetic nonlinearities in the SRM. Primarily, the variation of phase inductance with rotor position is expressed as a Fourier series wherein the first three components are considered.

The SRM drive system model includes the motor itself, the appropriate power electronic converter, and the related control system. The power electronic converter and the control circuitry are modeled based on their actual configuration. A typical representation of a classic converter for the SRM drive is shown in Figure 9.13. This system is an asymmetric driver with a converter with two switches per phase. Therefore, this driver is also named the two-switch-per-phase SRM driver.

In order to understand the working of the above converter, consider only one phase of the SRM (phase A). Turning on transistors T1 and T2 will circulate a current in phase A of the SRM. If the current rises above the commanded value, T1 and T2 are turned off. The energy stored in the motor winding of phase A will keep the current in the same direction (the current decreases). Hence, diodes D1 and D2 will become forward biased, leading to recharging of the source. This will decrease the current, rapidly bringing it below the commanded value.

In order to simulate the dynamic characteristics of the SRM, it is necessary to express the inductance–current–position characteristics of the motor accurately. As was mentioned earlier, for this purpose, the variation of the phase inductance with rotor position is expressed as a Fourier series expansion with only the first three

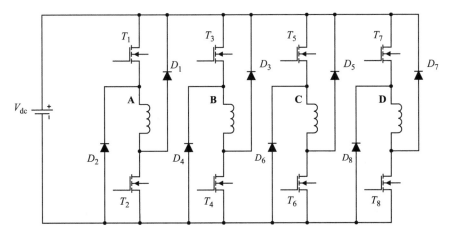

FIGURE 9.13 Layout of a typical classic (two-switch-per-phase) converter for an 8/6 SRM drive system.

terms being considered for analysis. In order to find these coefficients, three distinct points on the inductance–current waveform are
used. These points are inductance at the aligned position L_a,
unaligned position L_u, and at a position midway between the two
L_m. It is worth mentioning here that the mutual inductances of the
phases are neglected in this case. The voltage equation for the
conducting phase is given as follows:

$$v = Ri + l\frac{di}{dt} + \frac{d\psi}{dt}$$

where v is the voltage applied across the winding, R is the phase
winding resistance, l is the phase leakage inductance, $\psi = L(i,\theta)i$ is
the flux linkage, and L is the self-inductance of the phase.

As mentioned before, the self-inductance of one of the stator
phases is expressed by the first three terms in the Fourier series.
These three coefficients primarily depend on the current and can be
written as

$$L(i,\theta) = L_0(i) + L_1(i)\cos N_r\theta + L_2(i)\cos 2N_r$$

where

$$L_0 = \frac{1}{2}\left[\frac{1}{2}(L_a + L_u) + L_m\right]$$

$$L_1 = \frac{1}{2}(L_a - L_u)$$

$$L_2 = \frac{1}{2}\left[\frac{1}{2}(L_a + L_u) - L_m\right]$$

where N_r is the number of rotor poles.

$$L_a = L(\theta = 0°) = \sum_{n=0}^{n=k} a_n i^n$$

L_a is the aligned position inductance as a function of phase current.
Similarly, the inductance midway between the aligned and un-

aligned positions as functions of phase current can be expressed as follows:

$$L_m = L\left(\theta = \frac{\pi}{2N_r}\right) = \sum_{n=0}^{n=k} b_n i^n$$

$$L_u = L\left(\theta = \frac{\pi}{N_r}\right)$$

It is worthwhile mentioning here that the inductance at the unaligned position is assumed to be independent of the phase current, which happens to be a fairly valid assumption. In addition, k is the degree of approximation.

The phase inductance expressions for the other phases are basically shifted and are similar. The coefficients a_n and b_n are determined by curve-fitting methods. This ensures that the obtained inductance profile matches the profile which would be obtained experimentally. Back EMF is expressed as

$$e = \frac{d\psi}{dt} = \frac{d\psi}{d\theta}\frac{d\theta}{dt} + \frac{d\psi}{di}\frac{di}{dt}$$

where i and θ are independent variables. Hence, the expression for back EMF can be written as follows:

$$e = \frac{d\psi}{dt} = \frac{d(Li)}{dt} = \frac{d(Li)}{d\theta}\frac{d\theta}{dt} + \frac{d(Li)}{di}\frac{di}{dt}$$

$$e = i\frac{dL}{d\theta}\omega + i\frac{dL}{di}\frac{di}{dt} + L\frac{di}{dt}$$

The expression for L is used in order to derive a solution in a closed form for the back EMF. This is written as follows:

$$e = -\frac{N_r}{2}i\omega[(L_a - L_u)\mathrm{sin}N_r\theta + (L_a + L_u - 2L_m)\mathrm{sin}2N_r\theta]$$

$$+ \frac{di}{dt}\left\{\frac{1}{2}\left[\frac{1}{2}(L_a^* + L_u) + L_m^*\right] + \frac{1}{2}(L_a^* - L_u)\mathrm{cos}N_r\theta\right.$$

$$\left. + \frac{1}{2}\left[\frac{1}{2}(L_a^* + L_u) - L_m^*\right]\mathrm{cos}2N_r\theta\right\}$$

where

$$L_a^* = \sum_{n=0}^{n=k} (n+1)a_n i^n$$

$$L_m^* = \sum_{n=0}^{n=k} (n+1)b_n i^n$$

In addition to the above derived set of equations, an expression for the developed electromagnetic torque can also be derived in the closed form:

$$T = -\frac{N_r i^2}{4}[(L_a^{**} - L_u)\sin N_r\theta + (L_a^{**} + L_u - 2L_m^{**})\sin 2N_r\theta]$$

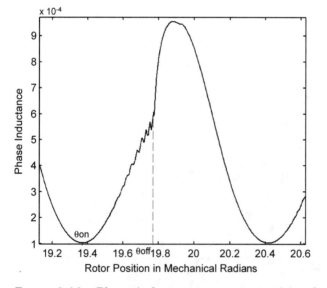

FIGURE 9.14 Phase inductance vs. rotor position from the dynamic modeling.

where

$$L_a^{**} = \sum_{n=0}^{n=k} \frac{2a_n}{n+2} i^n$$

$$L_m^{**} = \sum_{n=0}^{n=k} \frac{2b_n}{n+2} i^n$$

The mechanical equation of the drive system is given as follows:

$$T - T_L = J\frac{d\omega}{dt}$$

$$\omega = \frac{d\theta}{dt}$$

T_L is the load torque, which is generally a function of speed. Furthermore, J is the angular moment of inertia of all the rotating masses. The above equations present the dynamics of the SRM drive

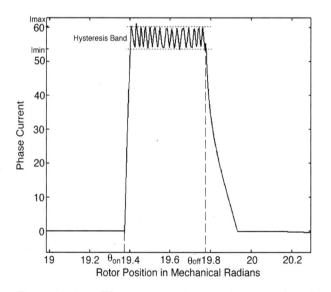

FIGURE 9.15 Phase current vs. rotor position from the dynamic modeling.

system. In addition, they also describe the terminal characteristics and the electromagnetic torque developed by one phase of the machine. As was mentioned earlier, the same applies to the other phases also, but with appropriate phase shifts and currents. It is important to note that the total electromagnetic torque developed by the motor is the sum of the instantaneous torques of the individual phases.

For a typical four-phase 8/6 SRM drive, the inductance model and phase current for phase 1 and the corresponding phase torque and radial force produced are given in Figures 9.14–9.18. These waveforms are based on the dynamic model of the SRM.

The inductance profile for all four phases of the 8/6 SRM is shown along with the phase currents, torque, and radial forces in

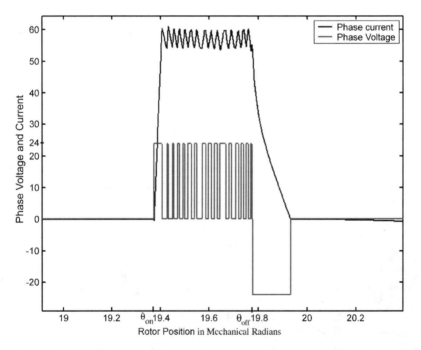

FIGURE 9.16 Phase voltage and current vs. rotor position from the dynamic modeling.

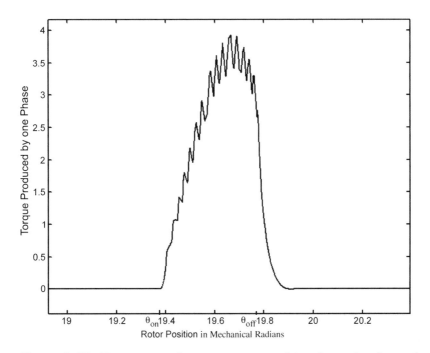

FIGURE 9.17 Torque per phase vs. rotor position from the dynamic modeling.

Figures 9.19–9.24. The current overlap provides higher torque output but introduces increased torque ripples. Figures 9.25 and 9.26 depict how torque varies with varying turn-on and turn-off angles.

The output torque and radial force with phase 2 under fault are shown in Figs. 9.27 and 9.28, respectively. It is found that the SRM operates close to a no-fault case but with reduced performance. This is primarily due to the absence of the excitation source on the rotor and the machine windings being physically and electromagnetically isolated from one another.

The simulation results indicate that maximum torque is obtained over −30 to −7 mechanical degrees, but is decreased when turn-on and turn-off angles are moved beyond thes regions. This was

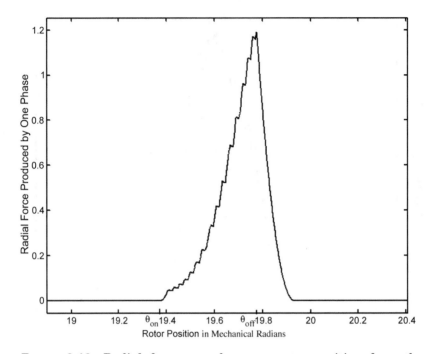

FIGURE 9.18 Radial force per phase vs. rotor position from the dynamic modeling.

also accompanied with higher torque ripples. Phase current overlapping provides the possibility of increasing the torque output by an appreciable degree without increasing the current levels, but again this brings high torque ripples and distorted radial forces. The output torque and radial force with phase 2 under fault highlight the potential of SRMs to operate reliably even under a phase fault with reduced performance.

9.5 CONTROL OF SRMs

The choice of the right control system is critical for the SRM system design. If the control strategy is defined properly, it provides a better

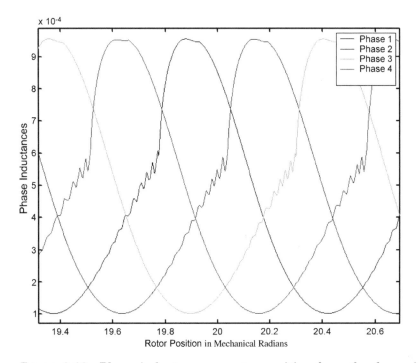

FIGURE 9.19 Phase inductance vs. rotor position from the dynamic modeling.

motor performance, lower energy usage, quieter operation, greater reliability, fewer system components, and a better dimension of the power elements.

Figure 9.29 shows phase inductance and ideal phase current and torque for the motoring and generating modes of operation. With the increase in inductance, the windings are excited for the motoring operation. The ideal waveforms of torque and current are shown in the figure for only one phase. However, by combining instantaneous values of the electromagnetic torque pluses for all phases, we can obtain total torque. It is seen that the machine produces discrete torque pulses, but by proper design of overlapping inductances we can produce continuous torque. For the motoring

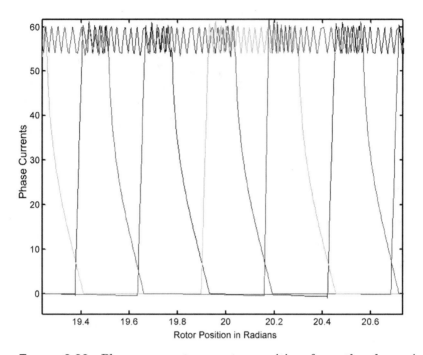

FIGURE 9.20 Phase current vs. rotor position from the dynamic modeling.

mode, in the ideal case, the current pulse is applied to the winding during the positive slope of the inductance (from the unaligned position to the aligned position). The average torque is controlled by adjusting the magnitude of the phase current or by varying the dwell angle (the angle from the turn-on instant to the turn-off instant). However, to reduce the torque ripples, it is advisable to keep the dwell angle constant and vary the magnitude of the phase current. By varying the dwell angle, we can ensure a safer operation with more control flexibility. In order to have a negative torque (generating mode), the current pulse is applied to the winding during the negative slope of the inductance (from the aligned position to the unaligned position).

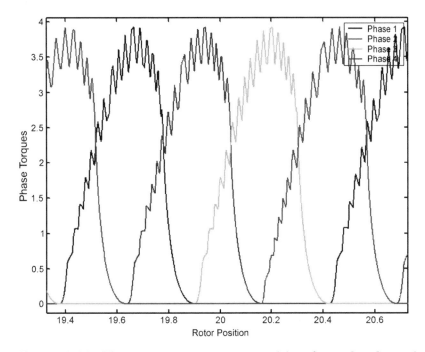

FIGURE 9.21 Phase torques vs. rotor position from the dynamic modeling.

9.5.1 Low-Speed Operation

The most popular control method for low-speed operation is the hysteresis control, or bang-bang control. This technique is also known as current control method. The switches are turned off and on according to whether the current is greater or less than the reference current. The error is used directly to control the states of switches. The phase current is limited within the preset hysteresis band. The waveforms are shown in Fig. 9.30. As the supply voltage is fixed, the switching frequency varies as the current error varies. Hence we can obtain precise current control. The tolerance band may be considered the main design parameter, but the noise filtering is difficult due to the varying switching frequency.

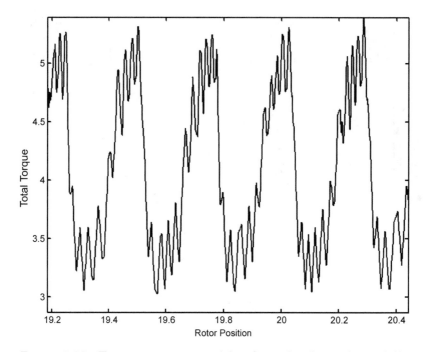

FIGURE 9.22 Torque vs. rotor position from the dynamic modeling.

Figure 9.31 shows the equivalent circuit of one phase of the classic two-switch-per-phase SRM driver. Figure 9.32 shows the equivalent circuit when the two switches are on. In this mode, the DC bus voltage is applied across the phase winding. Since the back EMF voltage is less than the DC bus voltage, the current increases. As soon as the current reaches the maximum current set by the hysteresis control, the two switches are turned off and, as a result, the two diodes conduct. Figure 9.33 shows the equivalent circuit when the two diodes are on. In this mode, the DC bus voltage is applied across the phase winding with negative polarity. Therefore, the phase current decreases. This mode continues until the phase current reaches the minimum current set by the hysteresis control (maximum current minus the hysteresis band). When the current reaches the minimum current, the two switches are turned on.

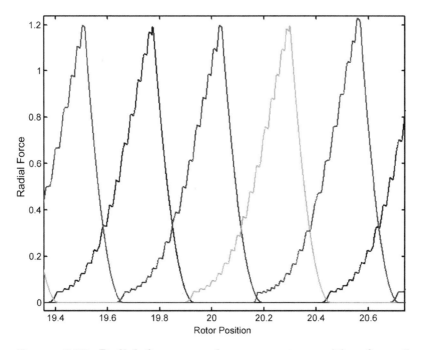

FIGURE 9.23 Radial force per phase vs. rotor position from the dynamic modeling.

Therefore, the two diodes are forced to be off. This hysteresis operation continues until the turn-off time. At the turn-off time, the two switches are turned off and the two diodes conduct. The current decreases until it reaches zero.

The technique for low-speed operation is a bipolar switching scheme because $+V_{DC}$ and $-V_{DC}$ are applied across the phase winding. Figure 9.34 shows the phase inductance, voltage, and current for the unipolar switching scheme. In this method, when the current reaches the maximum limit, instead of turning both switches off, only one of the switches is turned off. Therefore, one switch and one diode conduct and the voltage acros the phase winding is zero. The current decreases, but slower than in the bipolar method. The hysteresis operation continues until the turn-off

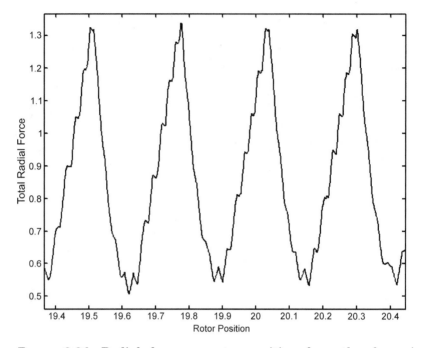

FIGURE 9.24 Radial force vs. rotor position from the dynamic modeling.

time when both switches are turned off. The two diodes conduct afterward and the voltage across the phase winding is $-V_{DC}$. The switching frequency in a unipolar switching scheme is less than the switching frequency in a bipolar switching scheme.

Bipolar and unipolar switching schemes have variable switching frequency. This is their main disadvantage. In order to have a constant switching frequency, a pulse width modulation (PWM) technique can be used. This strategy is useful in controlling currents at low speeds. Here the supply voltage is chopped at a fixed frequency with a duty cycle depending on the current error. Thus, both current and rate of change of current can be controlled. There are two main PWM techniques. In the first method, both switches are driven by the same pulsed signal, i.e., two of them are switched

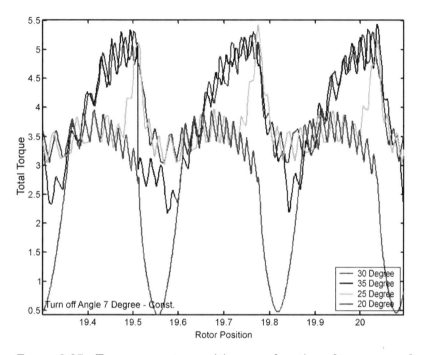

FIGURE 9.25 Torque vs. rotor position as a function of turn-on angle from the dynamic modeling.

on and off at the same time. This makes the overall design simple and cheap, but it increases current ripples. The second method is to keep the low-side switch on during the dwell angle, and the high-side switch is switched according to the pulse signals. This may help in current ripple minimization. Figure 9.35 shows the phase inductance, voltage, and current for the PWM switching technique.

9.5.2 High-Speed Operation

As the motor speed increases, it becomes difficult to regulate the current because of the combination of the back EMF effects and a reduced amount of time for the commutation control. At high

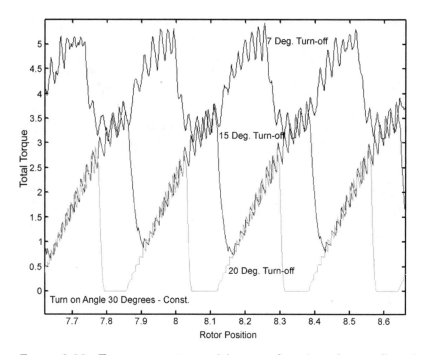

FIGURE 9.26 Torque vs. rotor position as a function of turn-off angle from the dynamic modeling.

speed, control can be obtained by increasing the conduction time (greater dwell angle), by advancing fire angles, or by controlling both. The difference between the turn-on angle and turn-off angle is known as the dwell angle.

Figure 9.36 shows the phase inductance, voltage, and current for high-speed operation. In high-speed operation, the current never reaches the hysteresis limit due to the larger back EMF. In fact, there is only one pulse for the current. Therefore, high-speed operation is also known as single-pulse operation. By adjusting the turn-on and turn-off angles so that the phase commutation begins sooner, we gain the advantage of producing current in the winding while the inductance is low and also of having additional time to reduce the

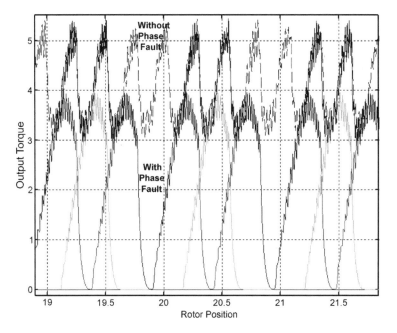

FIGURE 9.27 Output torque vs. rotor position with a fault in phase 2 from the dynamic modeling.

current in the winding before the rotor reaches the negative torque region. It should be noted that the marginal speed between the high-speed and low-speed regions is the base speed.

9.6 OTHER POWER ELECTRONIC DRIVERS

9.6.1 Miller Converter (*n* + 1 Topology)

Figure 9.37 shows the SRM driver with Miller converter. In this configuration, there is one switch for each phase. There is also one common switch for all phases. Therefore, this converter is also known as the one common switch configuration and *n* + 1 topology driver (*n* is the number of phases). This configuration has fewer

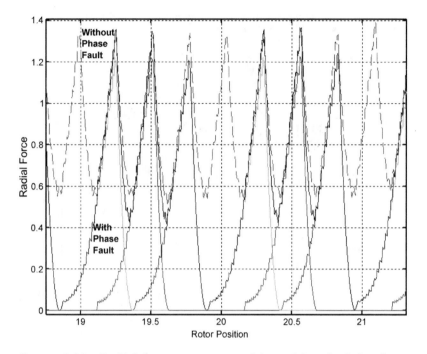

FIGURE 9.28 Radial force vs. rotor position with a fault in phase 2 from the dynamic modeling.

switches and diodes compared to the classic converter. Another advantages is that the Miller converter offers the lowest kVA inverter rating for a given drive power rating. However, the Miller converter has less flexibility. This is because the control angle range is limited since one switch is shared as the common switch. In addition, two phases of the machine cannot conduct simultaneously. Figure 9.38 shows the phase inductance, voltage, and current for a Miller converter.

A Miller converter has three modes of operation in low speed. In mode 1, as shown in Figure 9.39, S1 and S4 are turned on and D1 and D4 are off. Input DC bus voltage is applied across the phase winding. Therefore, phase current increases. Input source current is

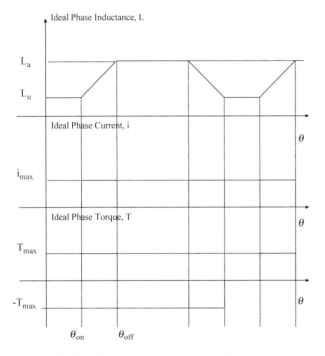

FIGURE 9.29 Phase inductance and ideal phase current and torque for motoring and generating modes of operation.

equal to the motor phase current. When the phase current reaches the maximum hysteresis current, S1 is turned off (mode 2 as shown in Figure 9.40). As a result, D4 conducts and the voltage across the phase winding is zero. In fact, this is a unipolar switching scheme. Input source current is zero. Phase current decreases until it reaches the minimum current set by the hysteresis control. Then S4 is turned on and the converter operates in mode 1. The converter switches between modes 1 and 2 until the turn-off time. At the turn-of instant, both S1 and S4 are turned off (mode 3 as shown in Figure 9.41). As a result, D1 and D4 conduct. Input DC bus voltage is applied across the phase winding with negative polarity. Therefore, phase current decreases until it reaches zero. During this mode, the input DC source is recharged.

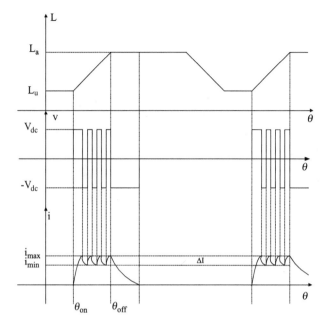

FIGURE 9.30 Phase inductance, voltage, and current for low-speed operation with hysteresis control.

9.6.2 R-Dump Converter

Figure 9.42 shows the SRM driver with an R-dump converter. In this configuration, there is only one switch for each phase without any common switch. Therefore, this is a low-cost SRM driver. However, the efficiency is low because of the dump resistor. Figure 9.43 shows the phase inductance, voltage, and current for an R-dump converter.

The R-dump converter has two modes of operation in low speed. In mode 1, as shown in Fig. 9.44, S1 is turned on. D1 is off since it is reverse biased because the capacitor voltage is applied across it with negative polarity. Input DC bus voltage is applied across the phase winding. Therefore, phase current increases. Input current is equal to the phase current. When the phase current

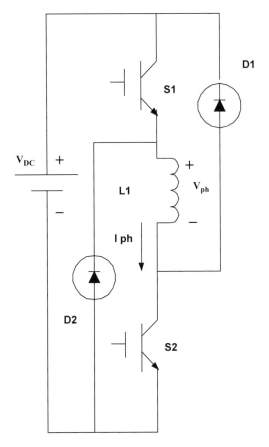

FIGURE 9.31 Equivalent circuit of one phase of a classic two-switch-per-phase SRM driver.

reaches the maximum hysteresis current, S1 is turned off (mode 2 as shown in Figure 9.45). D1 is forced to conduct. Since the capacitor voltage is greater than the DC bus voltage, a negative voltage is applied across the phase winding. Therefore, the phase current decreases. When the phase current reaches the minimum current set by the hysteresis control, mode 1 repeats. At turn-off time, the converter operates in mode 2 until the phase current reaches zero.

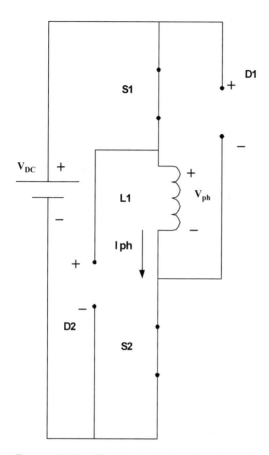

FIGURE 9.32 Equivalent circuit when the two switches are on.

Because of the resistive power loss in the dump resistor, in mode 2, efficiency of this driver is low.

9.6.3 C-Dump Converter

As shown in Fig. 9.46, in order to improve efficiency, instead of using a dump resistor like the R-dump converter, a C-dump converter uses an energy dump capacitor. This configuration is a

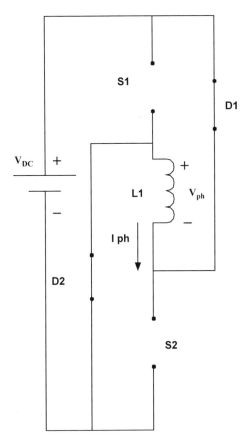

FIGURE 9.33 Equivalent circuit when the two switches are off.

single-switch-per-phase converter. Therefore, the packaging is compact. The DC source capacitor also has less ripple. In this topology, similar to the classic converter, both positive and negative DC bus voltage can be applied to the phase windings. This in turn provides control flexibility to improve the performance of the motor by reducing torque ripples and acoustic noise. However, the common switch has higher voltage and current ratings. Another disadvantage is that only motoring mode is possible. Figure 9.47 shows the phase inductance, voltage, and current for a C-dump converter.

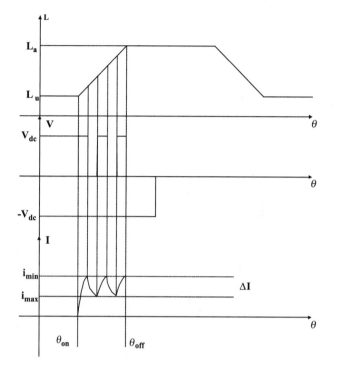

FIGURE 9.34 Phase inductance, voltage, and current for the unipolar switching.

The C-dump converter has seven operating modes. In mode 1, as shown in Figure 9.48, S1 is turned on; other switches and diodes are off. The input DC bus voltage is applied across the phase winding. Therefore, phase current increases. Input source current is equal to the phase winding current. When the phase current reaches the maximum hysteresis current, S1 is turned off. As a result, D1 is forced to conduct (mode 2, Figure 9.49). In mode 2, D4 and S4 are also off. The voltage applied across the phase winding is $-(V_c - V_{DC})$. Capacitor voltage is greater than the input DC bus voltage. The phase current decreases until it reaches the minimum hysteresis current. In mode 3, (Figure 9.50), all the switches and diodes are off. Phase voltage and current are zero.

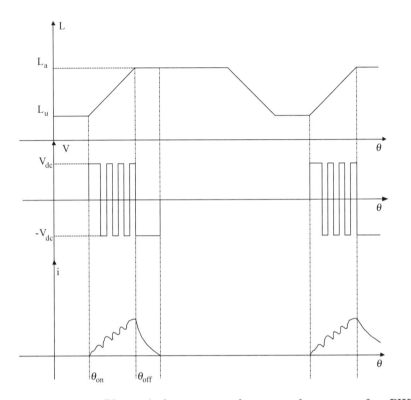

FIGURE 9.35 Phase inductance, voltage, and current for PWM switching.

In order to maintain a constant voltage across the dump capacitor, S4, D4, and L4 form a DC/DC buck converter. It is a step-down chopper if the dump capacitor and DC bus are considered input and output, respectively. It is also a step-up chopper if the DC bus and dump capacitor are considered input and output, respectively. The task of this DC/DC converter is to maintain the dump capacitor voltage. Figure 9.51 shows mode 4 when S4 is turned on while S1 is conducting. Figure 9.52 shows mode 5 when S4 is on while D1 is conducting. Both modes 4 and 5 can be used to charge the dump capacitor. However, common practice is to use only mode 4 for this purpose; therefore, mode 5 is not used as mode 4

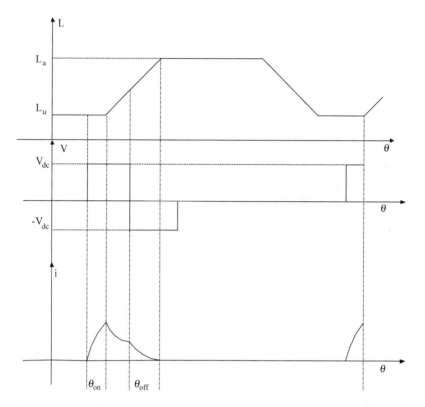

FIGURE 9.36   Phase inductance, voltage, and current for high-speed operation.

is enough to charge the dump capacitor. Simplicity is the advantage of mode 4 compared to mode 5.

In mode 6, D1 and D4 are on and S1 and S4 are off. In mode 7, S1 and D4 are on and D1 and S4 are off. In the most popular approach for the C-dump converter, only modes 1, 4, 6, and 2 are used consecutively.

9.6.4   Freewheeling C-Dump Converter

Figure 9.53 shows the SRM driver with freewheeling C-dump converter. Figure 9.54 depicts the phase inductance, voltage, and

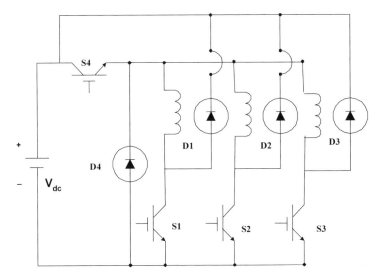

FIGURE 9.37 SRM driver with Miller converter ($n + 1$ topology).

current for this drive. This is also from the family of single-switch-per-phase converters. AC-dump converter cannot provide a zero voltage across the phase winding. Therefore, C-dump drivers always use bipolar switching schemes. However, a freewheeling C-dump converter can provide a zero voltage across the phase winding. As a result, a unipolar switching scheme as shown in Fig. 9.54 can be used. Reduced acoustic noise and torque ripples are the main advantages.

9.6.5 Split-DC Converter

Figure 9.55 shows the SRM driver with split-DC converter. Figure 9.6 depicts the phase inductance, voltage, and current for split-DC converter. In this configuration, a split-DC supply converter is used. The main disadvantage of this topology is derating the supply DC voltage. This is because only half the DC bus voltage is utilized as the applied voltage across the phase windings.

It should be noted that there are several other converter topologies for SRM drives. They include the buck-boost converter,

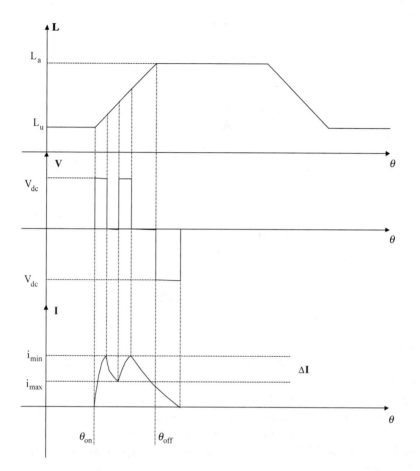

FIGURE 9.38 Phase inductance, voltage, and current for Miller converter.

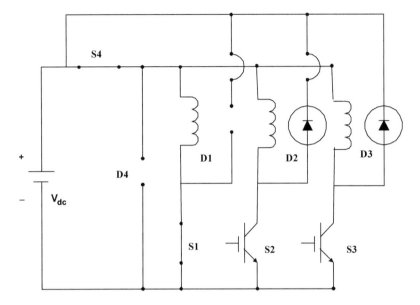

FIGURE 9.39 Equivalent circuit when S1 and S4 are on.

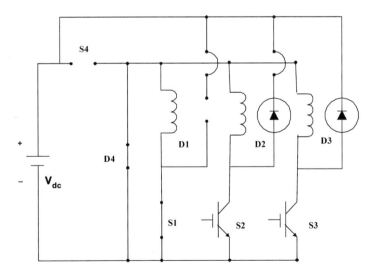

FIGURE 9.40 Equivalent circuit when S1 and D4 are on.

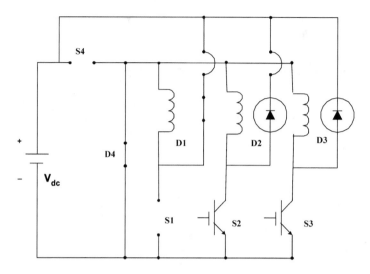

FIGURE 9.41 Equivalent circuit when D1 and D4 are on.

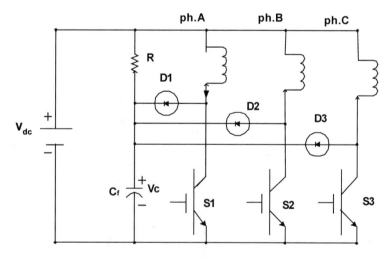

FIGURE 9.42 SRM driver with R-dump converter.

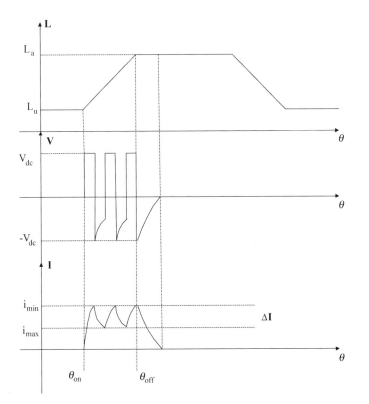

FIGURE 9.43 Phase inductance, voltage, and current for R-dump converter.

variable DC link voltage converters, the Sood converter, soft switched converters, and resonant converters.

9.7 ADVANTAGES AND DISADVANTAGES

Advantages of SRM drives are summarized as follows:

- There is saving in material cost as there is no winding or permanent magnet on the rotor.

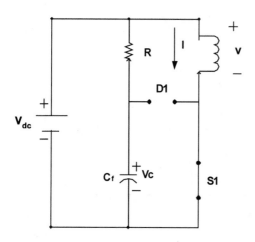

FIGURE 9.44 Equivalent circuit when S1 is on.

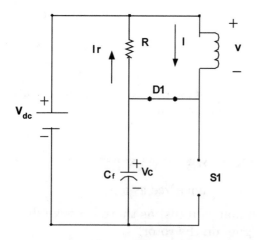

FIGURE 9.45 Equivalent circuit when S1 is off.

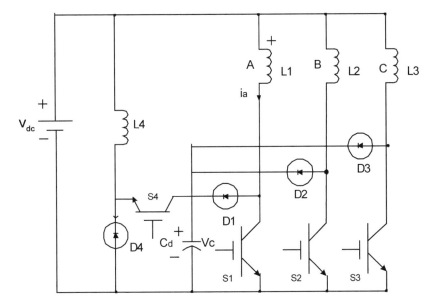

FIGURE 9.46 SRM driver with C-dump converter.

- Rotor losses are reduced due to the absence of rotor winding. The machine is therefore suited for low-voltage and current-intensive applications.
- Efficient cooling can be achieved as major losses are on the stator, which is easily accessible.
- As the size of the rotor is small, it has less moment of inertia, thus giving a large acceleration rate to the motor.
- Rotors are simple; hence, they are mechanically robust and therefore naturally suited for high-speed operation.
- The concentrated winding configuration reduces the overall cost compared to distributed windings. This configuration also reduces end turn buildup, which minimizes the inactive part of the materials, resulting in lower resistance and copper losses compared to the distributed windings in other machines.

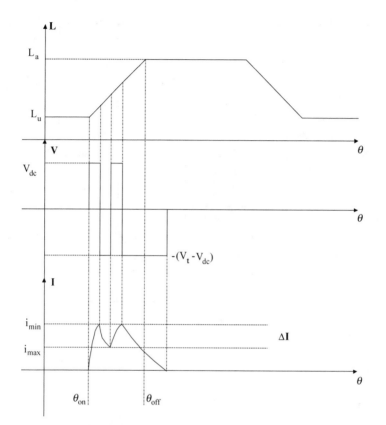

FIGURE 9.47 Phase inductance, voltage, and current for C-dump converter.

- As they are brushless machines, there is low maintenance cost.
- SRMs do not produce cogging or crawling torques; hence, skewing is not required.
- As windings are electrically separated from each other, they have negligible mutual coupling; hence, failure of one does not affect the other.
- SRMs show high reliability compared to other machines as there is freedom to choose the number of

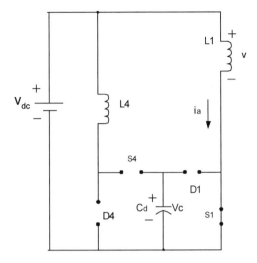

FIGURE 9.48 Equivalent circuit when S1 is on and other switches and diodes are off.

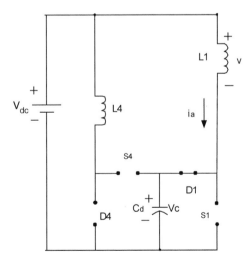

FIGURE 9.49 Equivalent circuit when S1 and S4 are off and D1 is on.

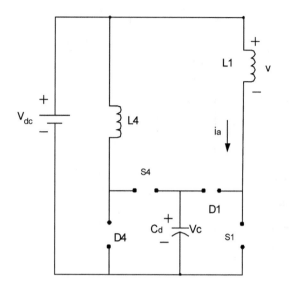

FIGURE 9.50 Equivalent circuit when all switches and diodes are off.

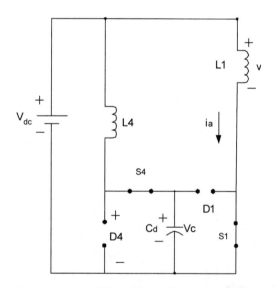

FIGURE 9.51 Equivalent circuit when S1 and S4 are on and D1 and D4 are off.

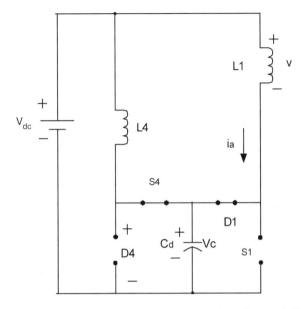

FIGURE 9.52 Equivalent circuit when D1 and S4 are on and S1 and D4 are off.

phases. As the number of phases increases, machine reliability also increases.

- They have inherent variable-speed operation.
- They provide a wide range of operating speed.
- SRMs offer great flexibility of control.
- Since power switches are in series with the phase windings and together they are parallel to the DC source voltage, there is no chance for shoot-through fault to occur; hence, higher reliability can be achieved.
- Their mechanical structure is not as stif as, say, synchronous machines and, coupled with the flexible control system, these machines are capable of effectively absorbing transient conditions, which provides resilience to the mechanical system.
- Extended speed constant-power operation.

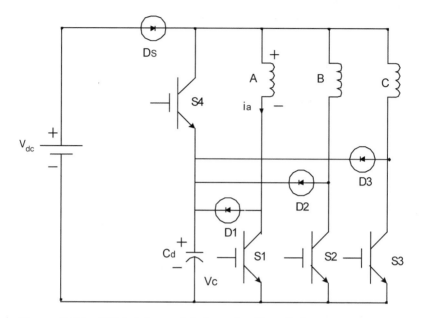

FIGURE 9.53 SRM driver with freewheeling C-dump converter.

Disadvantages of SRM drives are summarized as follows:

- High torque ripples are the main drawback of this machine, but controlling overlapping currents can reduce them.
- They exhibit high acoustic noise.
- Radial forces are minimal at unaligned positions and high at aligned positions; therefore, any variation over half rotor pitch can contribute to faster wear and tear of the bearings if there are rotor eccentricities and uneven air gaps, which is the major source of noise.
- Friction and windage losses are high due to the salient rotor.
- It lacks line start capability as it requires electronic power converter to run the machine.
- Position information is required to control SRMs; however, sensorless techniques can be used to avoid position sensors.

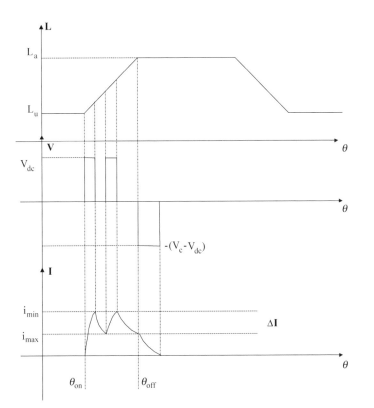

FIGURE 9.54 Phase inductance, voltage, and current for freewheeling C-dump converter.

- A separate freewheeling diode for each switch is necessary in all SRM converter topologies, which increases the overall cost compared to the H-bridge inverters.

SRM drives have many applications. Low-power applications include plotter drives, air handler motor drives, manual forklift/ pallet truck motor drivers, door actuator systems, air conditioners, and home appliances such as washers, dryers, and vacuum cleaners. Medium-power applications include industrial general purpose drives, train air conditioner drives, and mining drives. High-power

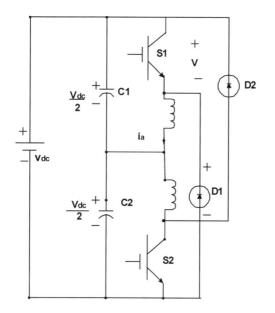

FIGURE 9.55 SRM driver with split-DC converter.

applications include electric propulsion systems, vehicular systems, domestic purpose applications like fans and pumps, and industrial adjustable-speed drives. There are also several high-speed applications such as screw rotary compressor drives, centrifuges for medical applications, and aerospace applications. Robust rotor construction and high power density are the advantages of the SRM drives for high-speed applications.

9.8 GENERATIVE MODE OF OPERATION

This is the operating mode of the switched reluctance generator (SRG) or the regenerative mode of a switched reluctance motor (SRM) when the drive is a two-quadrant or four-quadrant drive. In order to have a generating mode of operation, each phase is excited when the rotor is at the aligned position. In fact, the current pulse is provided during the negative slope of the phase inductance. Figures

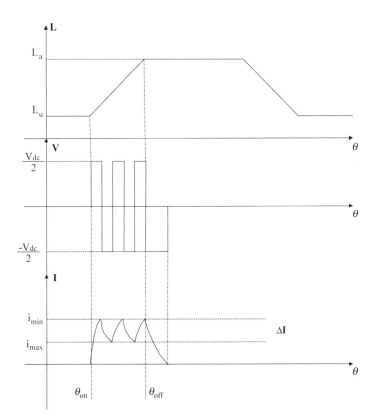

FIGURE 9.56 Phase inductance, voltage, and current for split-DC converter.

9.57 and 9.58 show the phase inductance, voltage, and current for low-speed (hysteresis or bang-bang control) and high-speed (single-pulse) operations, respectively.

The SRG has low radial vibrations as the stator phases are excited from the aligned to unaligned position; also there is absence of significant phase current. The presence of a large machine time constant also brings small radial forces. The use of antivibration configurations helps to further reduce the vibration. The radial forces being position dependent, the magnitude of the attractive forces is less and the absence of sudden changes in the rate of change

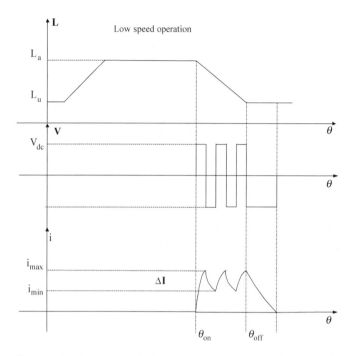

FIGURE 9.57 Phase inductance, voltage, and current for low-speed operation in generating mode.

of radial force in an SRG reduces the vibration to a large extent. Current profiling greatly reduces the noise level in an SRG, but it leads to lower performance levels.

The switched reluctance machine is unique in its operation as generator in that it does not require a permanent magnet or field windings on the rotor. The phase independence characteristic of the machine makes it extremely fault tolerant for critical applications. The absence of permanent magnets in these machines eliminates the problem of fire hazard of the machine generating into a shorted winding, compared to permanent magnet machines wherein one cannot turn off excitation. This unique feature makes the SRG a good candidate for various industrial and automotive applications.

A switched reluctance machine is operated in the generating mode by positioning phase current pulses during the periods where

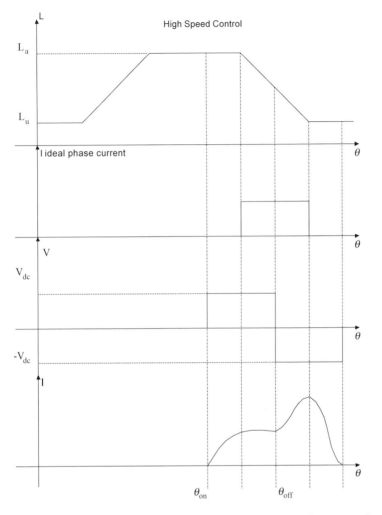

FIGURE 9.58 Phase inductance, ideal current, voltage, and current for high-speed operation in generating mode.

the rotor is positioned such that the phase inductance is decreasing. This occurs immediately after rotor and stator poles have passed alignment. Here, the machine obtains excitation from the same voltage source to which it generates power. Typically, a phase is turned on before a rotor pole aligns with that phase, drawing current from the DC source to excite the phase. Once the rotor poles pass alignment with the phase stator poles, the winding is disconnected from the DC source. It then generates into the same source with suitable connected diodes. The work done by the mechanical system to pull the rotor poles away from the stator poles is returned to the DC source. The DC source returns the excitation power plus the generated power. The control key is to position the phase current pulses to the phase stator poles in order to maximize the efficiency and to reduce the stresses on the power electronic switches. The phase current pulses during the generating mode are the mirror images of the phase current pulses in the motoring mode of operation.

An SRG is excited through a common asymmetric bridge. In a classic two-switch converter, one phase of this inverter uses the same DC source for exciting each SRG phase through two controllable switches and demagnetizing the same phase through the diodes. The conduction angle or dwell angle is the interval in which the phase is excited. Phase excitations depend on turn-on and turn-off angles. Invertors, switches, and diodes should be designed to support maximum DC source voltage and maximum phase current. Maximum voltage occurs when a switch reluctance machine operates as a generator. Main high-speed applications of SRGs are starter/alternators for automotive electrical systems and power generation in aerospace applications. The main low-speed application of SRG is for wind turbines.

9.9   ENERGY CONVERSION CYCLE

Two energy conversion cycles for an SRG at two different speeds are shown in Fig. 9.59 The area enclosed by the loop corresponds to the energy converted from mechanical to electrical form for the two cases. The dot in Figure 9.59 indicates the point where the control-

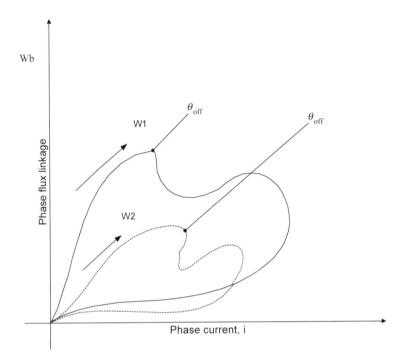

FIGURE 9.59 Energy conversion cycle.

lable switches are turned off and phase current is supported by diodes. Figure 9.60 gives the back EMF coefficient for the given value of the phase current. The back EMF coefficient is negative during the region of decreasing phase inductance and positive during the region of increasing phase inductance. For a certain level of the current value, the peak back EMF coefficient increases, but for any further increase in phase current, the back EMF coefficient decreases.

During excitation prior to the aligned position, phase current increases with back EMF, reducing effectiveness of the source voltage. This requires significant advancement in turn-on angle to have adequate phase current as the rotor enters the region of decreasing phase inductance. During demagnetization, phase current decreases if the source voltage is larger in magnitude than the

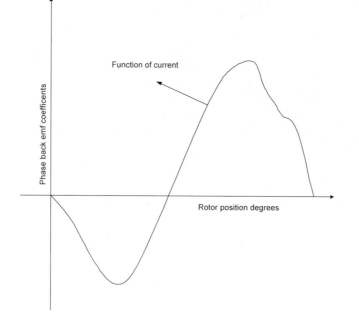

FIGURE 9.60 Phase back EMF coefficient vs. rotor position.

back EMF. At low-speed operation, multiple periods of excitations are required so that current is regulated to maintain adequate excitation as the rotor moves from the aligned position to the unaligned position. At high-speed operation, when switches are turned off, phase current increases first in the face of the negative source voltage and decreasing flux linkage.

SELECTED READINGS

1. Miller, T. J. E., Hendershot, J. R. (1993). *Switched Reluctance Motors and Their Controls*. Madison, WI: Magna Physics Publishing.
2. Krishnan, R. (2001). *Switched Reluctance Motor Drives: Modeling, Simulation, Analysis, Design, and Applications*. Boca Raton, FL: CRC Press.

3. Barnes, M., Pollock, C. (Nov. 1998). Power electronic converters for switched reluctance drives. *IEEE Trans. on Power Electronics*. 13:1100–1111.
4. Stefanovic, V. R., Vukosavic, S. (Nov./Dec. 1991). SRM inverter topologies: a comparative evaluation. *IEEE Trans. on Industry Applications*. 27:1034–1047.
5. Doo-Jin, Shin, Kyu-Dong, Kim, Uk-Youl, Huh. (2001). Application modified C-dump converter for industrial low voltage SRM. *Industrial Electronics Proceedings* 3:1804–1809.
6. Cheng, K. W. E., Sutanto, D., Tang, C. Y., Xue, X. D., Yeung, Y. P. B. (2000). Topology analysis of switched reluctance drives for electric vehicle. In: *Proc. 8th International Power Electronics and Variable Speed Drives Conf.*, pp. 512–517.
7. Emadi, A. (June 2001).Feasibility of power electronic converters for low-voltage (42V) SRM drives in mildly hybrid electric traction systems. In: *Proc. IEEE 2001 International Electric Machines and Drives Conference*. Cambridge, MA.
8. Le-Huy, H., Slimani, K., Viarouge, P. (Nov. 1990). A current-controlled quasi-resonant converter for switched reluctance motor. In: *Proc. 16th Annual Conf. of IEEE Industrial Electronics Society*. Vol. 2, pp. 1022–1028.
9. Uematsu, T., Hoft, R. G. (June 1995). Resonant power electronic control of switched reluctance motor for electric vehicle propulsion. In: *Proc. IEEE Power Electronics Specialists Conf.* Vol. 1, pp. 264–269.
10. Moallem, M., Ong, C. M. (Dec. 1990). Predicting the torque of a switched reluctance machine from its finite element field solution. *IEEE Trans. on Energy Conversion* 5(4):733–739.
11. Torrey, D. A. (Feb. 2002). Switched reluctance generators and their control. *IEEE Trans. on Industrial Electronics* 49(1):3–14.
12. Radun, A. (1994). Generation with the switched reluctance motor. In: *Proc. IEEE 9th Applied Power Electronics Conference and Exposition* pp. 41–47.
13. Patel, Y. P., Emadi, A. (Feb. 2003). Suitability of switched reluctance machines in distributed generation systems. In: *Proc. 2003 IASTED International Conference on Power and Energy Systems*. Palm Springs, CA.
14. Torrey, D. A., Lang, J. H. (Sept. 1990). Modeling a non-linear variable reluctance motor. *IEEE Proc. -Electric Power Applications*, 137(5):314–326.

15. Arkadan, A. A., Kielgas, B. W. (March 1994). Switched reluctance motor drive systems dynamic performance prediction and experimental verification. *IEEE Trans. on Energy Conversion* 9(1):36–43.

16. Franceschini, G., Pirani, S., Rinaldi, M., Tassoni, C. (Dec. 1991). Spice-assisted simulation of controlled electric drives: an application to switched reluctance motor drives. *IEEE Trans. on Industry Applications*, 27(6):1103–1110.

17. Fahimi, B., Suresh, G., Mahdavi, J., Ehsani, M. (1998). A new approach to model switched reluctance motor drive: application to analysis, design and control. In: *Proc. IEEE Power Electronics Specialists Conference.* Fukuoka.

18. Mahdavi, J., Suresh, G., Fahimi, B., Ehsani, M. (1997). Dynamic modeling of non-linear SRM using Pspice, In: *Proc. IEEE Industry Applications Society Annual Meeting.* New Orleans.

19. Fahimi, B., Suresh, G., Ehsani, M. (1999). Design considerations for switched reluctance motor: vibration and control issues. In: *Proc. IEEE Industry Application Society Annual Meeting.*

20. Lovatt, H. C., McClelland, M. L., Stephenson, J. M. (Sep. 1997). Comparative performance of singly salient reluctance, switched reluctance, and induction motors. In: *Proc. IEEE 8th International Electric Machines and Drives Conference,* pp. 361–365.

21. Wolff, J., Spath, H. (1997). Switched reluctance motor with 16 stator poles and 12 rotor teeth. In: *Proc. EPE '97.* Vol. 3. pp. 558–563.

22. Fahimi, B., Emadi, A., Sepe, R. B. (Dec. 2003). A switched reluctance machine based starter/alternator for more electric cars. *IEEE Trans. on Energy Conversion.* (in press).

23. Emadi, A. (Aug. 2001). Low-voltage switched reluctance machine based traction systems for lightly hybridized vehicles. *Society of Automotive Engineers (SAE) Journal,* SP-1633, Paper Number 2001-01-2507, pp. 41–47, 2001; and In: *Proc. SAE 2001 Future Transportation Technology Conference.* Costa Mesa, CA.

24. Husain, I. (Feb. 2002). Minimization of torque ripple in SRM drives. *IEEE Trans. on Industrial Electronics* 49:28–39.

25. Filizadeh, S., Safavian, L. S., Emadi, A. (Oct. 2002). Control of variable reluctance motors: a comparison between classical and Lyapunov-based fuzzy schemes. *Journal of Power Electronics* 305–311.

26. Fahimi, B., Suresh, G., Rahman, K. M., Ehsani, M. (1998).

Mitigation of acoustic noise and vibration in switched reluctance motor drives using neural network based current profiling. In: *Proc. IEEE Industry Application Society Annual Meeting.* St. Louis.

27. Cai, W., Pillay, P. (March 2001). Resonant frequencies and mode shapes of switched reluctance motors. *IEEE Trans. on Energy Conversion* 16(1):43–48.

28. Cai, W., Pillay, P. (May/June 1999). An investigation into vibration in switched reluctance motors. *IEEE Trans. on Industry Applications* 35(3):589–596.

29. Johnson, J. P., Rajarathnam, A. V., Toliyat, H. A., Suresh, G., Fahimi, B. (1996). Torque optimization for a SRM using winding function theory with a gap dividing surface. In: *Proc. IEEE Industry Applications Society Annual Meeting.* San Diego.

30. Rahman, K. M., Suresh, G., Fahimi, B., Rajarathnam, A. V., Ehsani, M. (May/June 2001). Optimized torque control of switched reluctance motor at all operating regimes using neural network. *IEEE Trans. on Industry Applications* 37(3):904–914.

31. Kjaer, P. C., Gribble, J. J., Miller, T. J. E. (Nov./Dec. 1997). High-grade control of switched reluctance machines. *IEEE Trans. on Industry Applications* 33(6).

32. Fahimi, B., Emadi, A. (June 2002). Robust position sensorless control of switched reluctance motor drives over the entire speed range. In: *Proc. IEEE 33rd 2002 Power Electronics Specialist Conference.* Cairns, Queensland, Australia.

33. Suresh, G., Fahimi, B., Rahman, K. M., Ehsani, M. (June 1999). Inductance based position encoding for sensorless SRM drives. In: *Proc. 30th Annual IEEE Power Electronics Specialists Conf.* Vol. 2. South Carolina, pp. 832–837.

10

Utility Interface Issues

Over the last decade, the electric machine industry has been placing more emphasis on variable- and adjustable-speed operations. Developments in power electronics made it possible to achieve these goals. However, this technology produces pollution to the utility grid and raises power quality issues. This chapter describes recent harmonic regulations imposed on power electronic equipments and systems. Merits and demerits of poor power quality drives are also explained. As a solution, various power factor correction (PFC) techniques are introduced for advanced motor drives. Different techniques for active and passive filters, special drive topologies for PFC, and harmonic injection are explained in detail.

Most traditional motor drives are fixed-speed drives. Recent advancements in the power electronic industry have encouraged machine manufacturers to start developing systems for adjustable- or variable-speed/frequency drives. Among the reasons for adopting adjustable-speed drives (ASDs) are better overall performance, improved efficiency, and reduced hardware complexity. Power electronics is used as an interface between the utility and motor. In other words, the machine is driven by a power electronic interface rather than the utility supply. The obvious reason is to achieve a variable form of supply, which is not available from the utility. By adjusting this variable supply, machine performance can be optimized. A typical advanced electric motor drive architecture is shown in Fig. 10.1. In the figure, the utility supply can be single phase or three phase. The machine can be single phase, three phase, or even multiphase, depending on the performance requirement. The power electronic system is designed in a way such that the machine can be optimized for its own performance independent of the available supply. Sensors can be Hall-effect or it is also possible to achieve sensorless drives.

The increased use of ASDs in power systems has led to a major problem for power quality. ASDs along with their power electronic drives appear as nonlinear loads to the utility grid or power system. In general, any power electronic driven system im-

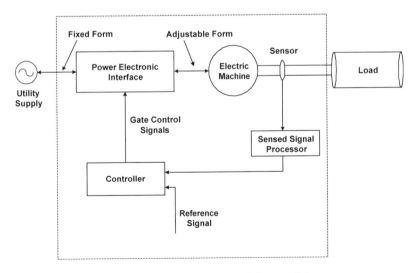

FIGURE 10.1 Advanced electric motor drive architecture.

poses power quality issues, i.e., power factor deterioration and harmonic distortion. The solution to these problems also resides in power electronics. Figure 10.2 shows the input current of a typical ASD. The harmonics of orders 5 and 7 are considerably high. Any electric motor drive system can be simplified, as shown in Fig. 10.3.

If the load is linear or the machine is directly connected to the mains, the supply voltage and current waveforms are sinusoidal and the power factor is given as follows:

$$\cos\Phi = \frac{V_s}{I_s}$$

However, advanced electric motor drives usually consist of a diode bridge rectifier followed by a bulk capacitor and the power processing stage. The bulk capacitor is used to smooth the output

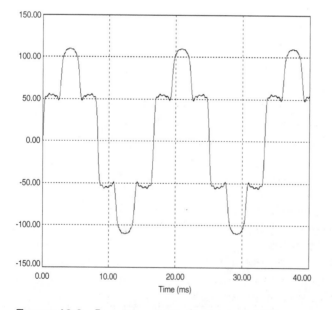

FIGURE 10.2 Input current of a typical ASD.

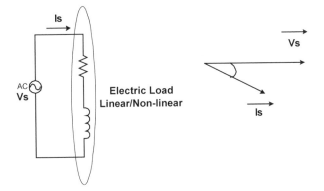

FIGURE 10.3 Simplified electric motor drive system.

voltage waveform. The diode network and the capacitor draw current from the mains only when the instantaneous mains voltage is greater than the capacitor voltage. This results in a nonsinusoidal supply current waveform. In this case, the power factor is not given by the above simple equation. Now the power factor is defined as

$$PF = \frac{DPF}{\sqrt{1 + THD^2}}$$

where DPF is the displacement power factor and THD is the total harmonic distortion. DPF and THD are defined as follows:

$$DPF = \cos\Phi_1 = \frac{V_s}{I_{s1}}$$

$$THD = \frac{I_{distortion}}{I_{s1}}$$

where I_{s1} is the fundamental component of the supply current. Since the capacitor is chosen for a certain hold-up time, its time constant is much greater than the frequency of the mains. This implies that the instantaneous mains voltage is greater than the capacitor voltage only for very short periods of time (charging time of capacitor). During this time the capacitor must charge fully. Therefore, large pulses of current are drawn from the line over very short periods of time. This is true of all rectified AC sine wave signals with capacitive filtering. This causes the following problems:

- Creation of harmonics and electromagnetic interference (EMI)
- High losses
- Required overdimensioning of parts
- Reduced maximum power capability from the line

Power factor correction makes the load look more like a resistive element than would be the case without PFC. Modern PFC circuits can achieve power factors very near to unity. PFC circuits have the following advantages:

- Better source efficiency
- Overall lower power installation cost
- Lower conducted EMI
- Reduced peak current levels
- Act as filters for the conducted EMI
- Make possible common input filtering for paralleled supplies because the loads all appear to be resistive
- Better chance of agency approval

However, PFC circuits have one or more of the following disadvantages:

- Introduce greater complexity into the design
- Have more parts, adversely affecting reliability and cost
- Generation of EMI and radiofrequency interference (RFI) requires extra filtering, making the input filter more complex and more expensive
- Higher system cost

Harmonic standards developed by IEEE and IEC are enforced in many parts of the world including Europe. Therefore, PFCs have attracted a lot of attention in the power electronic industry. In most applications, it is not difficult to meet these standards; however, the most economic choices are still being developed. IEEE and international harmonic standards can be grouped into three main categories:

1. Customer system limits:
   - IEEE 519-1992
   - IEC 1000-2-2 (compatibility levels)
   - IEC 1000-3-6

2. Equipment limits:
   - IEC 1000-3-2
   - IEC 1000-3-4
   - New task force in IEEE (harmonic limits for single-phase loads)

3. How to measure harmonics
   - IEC 1000-4-7

The IEC 1000 series deals with electromagnetic compliance. Part 3 sets limits and Series 2 addresses limiting harmonic current emission for equipment input current less than or equal to 16 A. IEC 1000-3-4 not only deals with individual equipment, but also sets limits for the whole system installation. Both single-phase and three-phase harmonic limits are addressed in this section of the regulation. On the other hand, IEEE standard 519 sets limits of harmonic voltage and current at the point of common coupling (PCC). The philosophy behind this standard is to prevent harmonic currents traveling back to the power system and affecting other customers.

10.1 ASD EXAMPLE

As an example, a C-dump converter for an advanced switched reluctance motor (SRM) drive is investigated. The SRM cannot be operated by directly connecting to the mains. A power electronic interface is absolutely necessary for running it optimally. Various power electronic topologies have been proposed. C-dump converter topology is one of the popular topologies to drive SRMs. As this topology contains maximum number of passive elements and complexity, it is chosen to verify the PFC function for advanced drives as a worst-case scenario. A power stage diagram and operating modes circuit reductions are shown in Figs. 10.4 and 10.5.

Three inductors grouped together represent three phases of the machine. The C-dump converter is operated in such a way that the required phase is energized and de-energized at required instances, i.e., aligned and unaligned positions of the rotor. Both hysteresis current control and PWM control are possible. Performance of the PFC circuit remains the same in all operating conditions. Even with different drives of different machines, this general PFC approach is equally effective.

The switches are turned on when the rotor is at the unaligned position, and they should not be conducting after the rotor phase passes the aligned position with the stator. There are various modes possible depending on particular applications. Here, when switches are turned on, the phase inductor and dump inductor store energy, and when the switches are turned off, both of them release the energy. The dump capacitor charges in a half cycle and discharges in

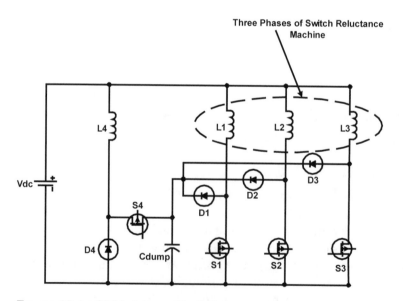

FIGURE 10.4   SRM drive with C-dump converter.

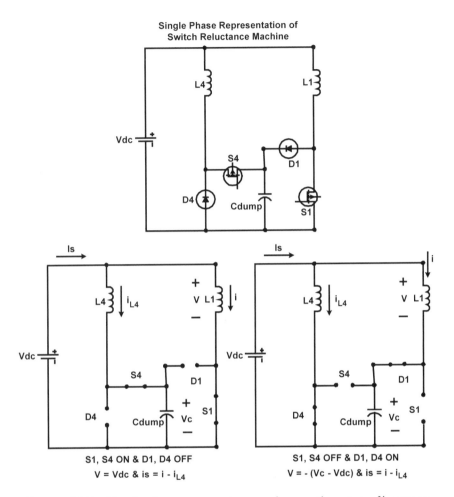

FIGURE 10.5 Single-phase power stage and operating state diagrams of C-dump SRM drive.

another half cycle. The voltage level of this capacitor should be higher than the DC link voltage. Figure 10.6 shows the phase voltage and current waveforms when the drive is operated with the DC supply.

However, this is not the case with the home or commercial applications of many drive systems. Those drive systems pass

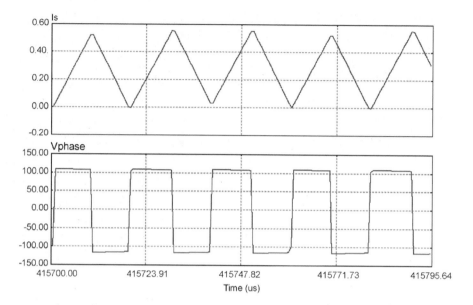

FIGURE 10.6 Phase current and voltage of SRM with DC supply at the front end.

through one or more power conversion stages. In case of advanced DC drive systems, the AC supply of the utility is usually converted using a simple diode bridge rectifier (DBR). As mentioned before, when only a DBR is connected between the drive and utility, the smoothing capacitor gets charged and discharged during the high line periods (short time intervals), and high current spikes occur. This deteriorates both power factor and overall system performance. Figure 10.7 explains the power stage of such uncontrolled DC link. Waveforms are depicted in Fig. 10.8. From the supply voltage and current waveforms shown in Fig. 10.8, it is clear that high distortion occurs on the supply side.

The effects of poor power factor are illustrated in this section. This has encouraged various organization to introduce standards for allowable harmonics and minimum power factors of power electronic systems. By doing so, power quality of the overall power system can be improved by a considerable amount. In addition, ma-

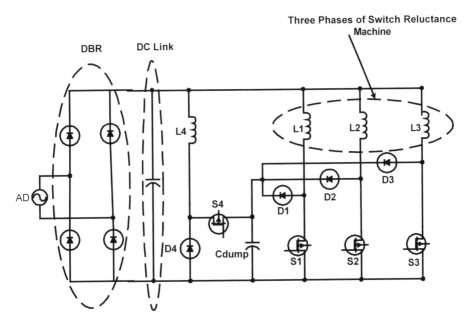

FIGURE 10.7 SRM drive with DBR at its front end as uncontrolled DC link.

terial cost saving can be realized in home appliance systems. The next section discusses different basic methods of power factor correction.

10.2 POWER FACTOR CORRECTION METHODS

Depending on the circuit elements employed in the power factor correction technique, power factor correction can be subdivided into passive PFC, active PFC, and harmonic injection. Some time-special topologies are to be employed to reduce the component count. All of these techniques are explained here with simulation examples.

Passive PFC This correction method is the simplest and is employed by most power processor circuits, such as switching power supplies and advanced motor drives. It is very effective in lowering

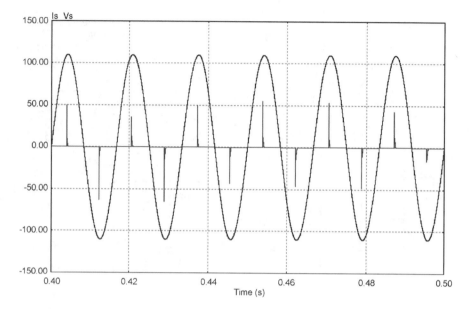

FIGURE 10.8 Input voltage and current of the SRM driver with DBR.

and limiting current harmonic distortions. By proper design, it is also effective in achieving good power factors on the line side. These passive PFC circuits, as passive filters, can be added to either the AC or DC side. Generally, they are placed between the diode bridge rectifier mentioned before and DC link capacitor, i.e., on the DC side. Figures 10.9 and 10.10 show the passive PFC circuit and its effect on the supply line and voltage waveforms in simulation results.

As mentioned in the previous section, the nonlinear load power factor depends on displacement power factor too. Generally, a capacitor is added on the AC side to improve DPF. It can be seen from Fig. 10.10 that both phase current and voltage are not changed, but the supply current is very much in phase with the supply voltage. In other words, the power factor is improved considerably. In addition, peak of the input side current is reduced a lot. Thus, THD is also reduced to a great extent.

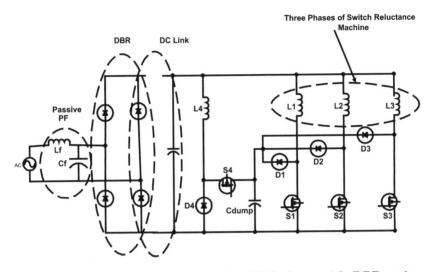

FIGURE 10.9 SRM drive with passive PFC along with DBR at its front end.

It is observed that the power factor with this method is not improved very close to unity. The size of the passive elements, i.e., inductor and capacitor, is also an issue. When more than one conversion stage is involved in the application, relative stability and EMI are the factors to be kept in mind while designing and inserting passive PFC circuits, specifically magnetic components, in the system. However, low cost and simple implementation make this technique attractive to many manufacturers.

Active PFC   As this technique involves active switching elements, it is known as active PFC. There are many ways of achieving active power factor correction. A boost converter is the most popular and simplest one. Figures 10.11 and 10.12 show the power stage diagram and simulation results for an SRM drive with boost converter.

From the waveforms shown in Fig. 10.12, it is clear that by using a DC/DC boost converter, a power factor of almost unity can be achieved, and the load to the converter appears almost resistive. No current peak is visible in the supply current waveform. Here, a boost converter switch is controlled, keeping output voltage of the

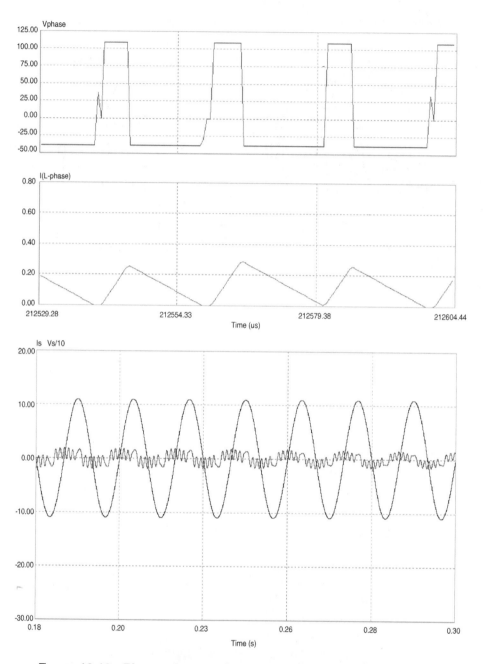

FIGURE 10.10 Phase voltage and current waveforms and supply voltage and current waveforms with passive PFC.

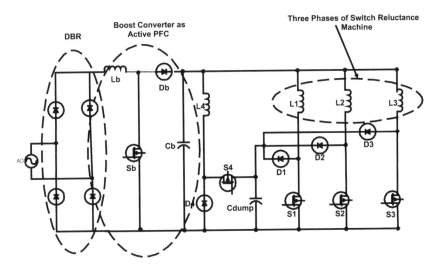

FIGURE 10.11 SRM drive with boost converter along with DBR at its front end.

converter regulated, but this has nothing to do with controlling the switches of the main SRM drive circuit. During dynamic conditions, it should be observed that the overall system is not becoming unstable. Sometimes all the switches in a system are synchronized to avoid this problem. By modulating the duty cycle of the boost converter switch, the input current can be controlled to track the input voltage. With low distortion and accurate tracking between current and voltage, the power factor obtained from adding a front-end boost converter is typically higher than 0.99, and the input current THD is normally less than 5%. Limitation of the boost converter is that the output voltage should be always greater than the maximum peak supply voltage. To alleviate this problem, another PFC circuit has been developed called the buck-boost converter (Figs. 10.13 and 10.14). This active PFC topology can deliver output voltage both less and greater than the supply voltage depending on the line and load situation. Input voltage can also be tracked by controlling the current in a specific manner. Figures 10.13 and 10.14 show a buck-boost converter at the front end of the SRM drive and its simulation results.

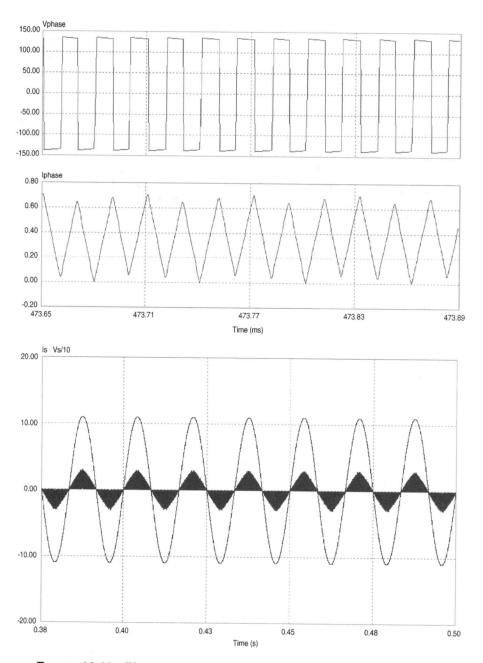

FIGURE 10.12 Phase voltage and current waveforms and supply voltage and current waveforms.

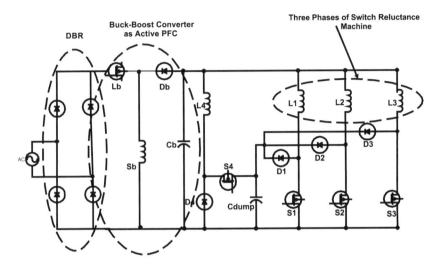

FIGURE 10.13 Buck-boost converter at the front end of the SRM drive.

The boost and buck-boost converters seem to be overcompensating, if satisfying the standards is the only concern. The issue of choosing a passive or active PFC seems to be the tradeoff between cost and effectiveness. The effectiveness means the extent that harmonics are eliminated or reduced, not how well the method complies with the standard. However, with the trend of continuous reduction in semiconductor cost, this barrier will also soon be removed.

Specifically for the SRM, to achieve a high power factor, changes are proposed in the C-dump converter (Fig. 10.15). It has been proved that a very high power factor can be achieved using this SRM drive. This is almost the same C-dump converter discussed earlier with some changes. This also falls in the category of active PFC, or one might say it is a hybrid PFC technique. This is because the switches in the drive circuits are controlled to achieve not only the required phase current and voltage, but also a very high power factor. It can also be seen from the power stage diagram shown in Fig. 10.15 that the mutual inductor just before the DC link capacitor appears almost identical to the passive PFC circuit shown

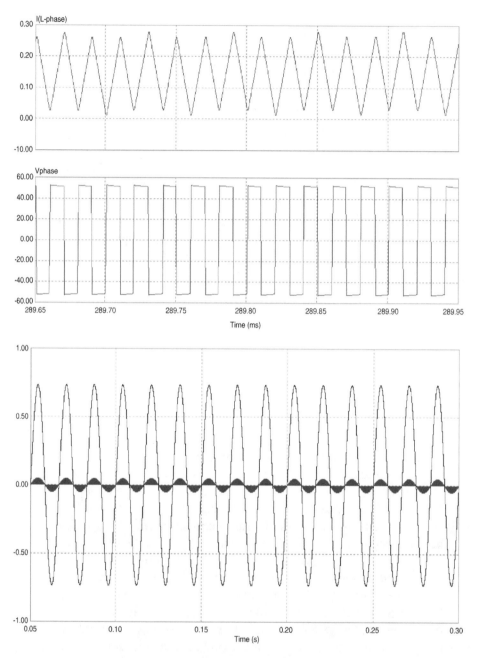

FIGURE 10.14 Phase voltage and current waveforms and supply voltage and current waveforms.

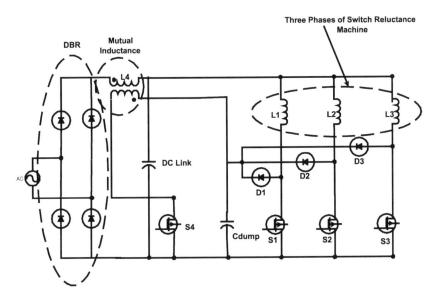

FIGURE 10.15 SRM driver with modified C-dump converter.

previously in this section. Figures 10.16 and 10.17 show the operating modes and simulation results, respectively, of the same converter. A very high power factor without noticeable current distortion can be observed from the simulation results.

Harmonic Injection/External Compensation This method is generally used to filter the harmonics from the line but not to improve the power factor. However, it can be employed for power factor correction. The sizes of passive components increase when they have to remove low harmonic components (source current and voltage are low-frequency signals). This filter is usually configured to plug into an outlet and serve as a plug-in point for 2-to-4 electronic devices. There are three main types of this filter: parallel resonant, series resonant, and series-parallel resonant.

10.3 ACTIVE POWER FILTERS

With the increase of nonlinear loads such as ASDs drawing nonsinusoidal currents, power quality distortion has become a serious

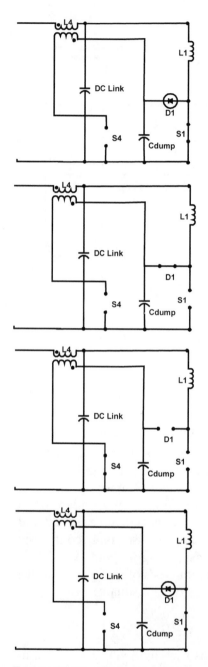

Figure 10.16 Operating modes of the modified C-dump converter.

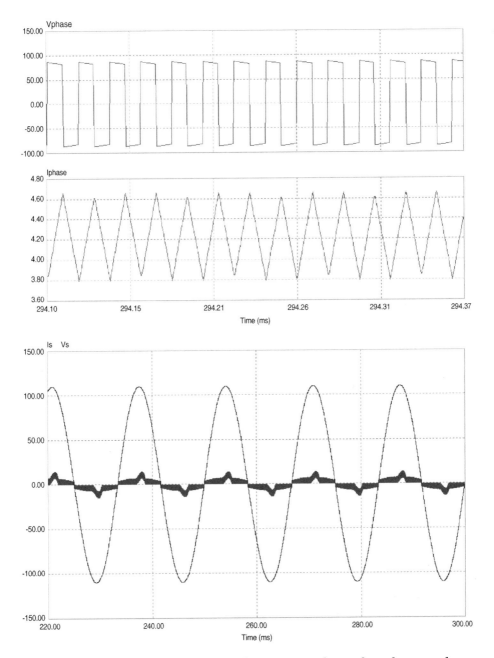

FIGURE 10.17 Phase voltage and current and supply voltage and current waveforms.

problem in power systems. Active filters (AFs) are used for harmonic mitigation as well as reactive power compensation, load balancing, voltage regulation, and voltage flicker compensation. Based on topology, there are two kinds of active filters: current source and voltage source. Current-source active filters (CSAFs) employ an inductor as the DC energy storage device. In voltage-source active filters (VSAFs), a capacitor acts as the energy storage element. VSAFs are cheaper, lighter, and easier to control compared to CSAFs. There are also four types of active filters based on the system configuration.

Current-source active filters use a current-source inductor. This type of energy source is commonly used in shunt-type active filters. Different configurations for this kind of active filter have been developed in forms of low-power, single-phase or high-power, three-phase three-wire or four-wire systems. In the three-phase, four-wire system, load current unbalance can be compensated in addition to current harmonics and reactive current.

In CSAFs, the DC current of the energy storage inductor must be greater than the maximum harmonic of the load (maximum deviation of source current from reference value). If the current of the DC inductor is too small, the inverter cannot do proper compensation. This DC current should not be too much. If the current is too much, excessive loss results in the inductor and inverter; a passive filter cannot cancel switching frequency. There is no need for the DC power supply because an active filter only delivers reactive power and a small amount of fundamental current needed to compensate the AF losses.

A small capacitor is used to protect switches against overvoltages and also to make a low-pass LC filter with the inductor between the active filter and system to suppress switching frequency. For preventing resonance, the resonance frequency of the passive filter must be greater than the highest frequency of harmonics and considerably less than the switching frequency. The control strategy must be well designed to prevent this resonance.

The most dominant type of active filter is the voltage-source inverter (VSI) active filter. Their design has been improved and they have been used for many years; now they are at the commercial stage. They are lighter, cheaper, and easier to control compared to the current-source inverter (CSI) type. Their losses are less than CSAFs, and they can be made in multilevels and multisteps.

Voltage-source active filters employ a capacitor as the DC energy storage. They are presented in single-phase or three-phase, three-wire or four-wire systems. This kind of active filter is convenient in uninterruptible power supply (UPS) systems. In UPS systems, DC energy storage is available and a DC/AC inverter is also ready. Only a control strategy is needed to convert the UPS to an AF when the source is in normal condition. Different kinds of control techniques are used to control VSAFs. The well-known control techniques are the instantaneous d-q theory, synchronous d-q reference frame method, and synchronous detection method.

In VASF, the DC voltage of the energy storage capacitor must be greater than the maximum line voltage. For proper operation of the active filter, at any instant the voltage of the DC capacitor should be 1.5 times of the line maximum voltage. A linking inductor establishes a link between the filter and system. The AF delivers its current to the system through the inductor. For controllability of AF, this inductor should not be large.

Active filters can also be classified as shunt, series, and hybrid. The most popular type of AF is the shunt type. Shunt AFs can be single-phase or three-phase, VSI or CSI. Shunt AFs are used to compensate the current and voltage harmonics of nonlinear loads, to perform reactive power compensation, and to balance unbalance currents. A shunt AF senses the load current and injects an appropriate current into the system based on its control function. Shunt AFs are currently commercially available.

A shunt AF acts as a current source. The sum of its current and load current is the total current, which flows through the source. Therefore, controlling the output current of an AF can control the source current. Ratings of series and shunt AFs have been compared in some papers. Based on those studies, the shunt AFs has approximately half the switch power rating of series AFs. The peak voltage over switches in series AFs is about one-third the peak voltage over switches in shunt AFs.

Series AFs can also be single-phase or three-phase and employ voltage-source or current-source inverters. Series AFs are mostly used to compensate voltage harmonics produced by nonlinear loads as well as voltage regulation and voltage unbalance compensation.

Series AFs are located in series between source and nonlinear loads. In the presence of source-side impedance, voltage harmonics of the nonlinear load appear at the point of common coupling. Series AFs sense the load-side voltage and produce the harmonic of load voltage in the negative direction and makes the voltage at the point of common coupling free of harmonics.

The main purpose of using a hybrid of active and passive filters is reducing the initial cost of the filter and improving the efficiency. Many configurations and combinations of active and passive filters have been studied and developed. Experimental results of combination series and shunt AFs with shunt passive filters are presented in many papers. Usually, the passive filter is tuned to specific frequency to suppress that frequency, decreasing the power rating of AF. Shunt passive filters should also be high pass to cancel the switching frequency of the AF and high-frequency harmonics. In this case, the switching frequency of the AF will decrease.

Another problem which AFs are faced with is high fundamental current through series AFs and high fundamental voltage across shunt AFs. Paralleling of series AFs with a passive filter can solve high current problems in series AFs. A proper control strategy should be adopted to avoid the possibility of resonance. High voltage across shunt AFs is reduced by putting the shunt AF in series with a passive filter.

Unified power quality conditioners (UPQCs), also known as universal AFs, are ideal devices to improve power quality. A combination of series and shunt AFs forms the UPQC. Series AFs suppress and isolate voltage harmonics, and shunt AFs cancels current harmonics. Usually, the energy storage device is shared between two AFs, either in CSI or VSI. There are two kinds of UPQC. In the first type, a shunt AF is placed near the source and a series AF is placed near the load. The series AF is used to compensate voltage harmonics of the load and the shunt AF is used to compensate residual current harmonics and improve power factor or to balance the unbalanced load. In the second type, a shunt AF is placed near the load to compensate current harmonics of load, and a series AF is placed near the source to compensate voltage harmonics of the source or regulate the voltage.

In conclusion, IEEE and other international standards are imposing limits on harmonic voltages and currents. Many power electronic circuit designs have been proposed to deal with these standards. Effectiveness of active PFC is normally not a problem, but the cost involved in the additional power electronic circuit could be a major obstacle to acceptance. The simplest power factor correction method is to use passive LC filters to comply with IEC and IEEE standards. Although these passive PFC methods comply with the standards, the problems of EMI, EMC, and size of the passive elements involved must be addressed. Thus, the design of cost-effective power electronic equipment that complies with harmonic standards without introducing side effects or system interaction problems remains an open challenge to power electronic and motor drive engineers.

SELECTED READINGS

1. Dugan, R. C., McGranaghan, M. F., Beaty, H. W. (1996). *Electrical Power Systems Quality*. New York: McGraw-Hill.
2. Bollen, M. H. J. (2000). *Understanding Power Quality Problems: Voltage Sags and Interruptions*. Piscataway, NJ: IEEE Press.
3. Arrillaga, J., Watson, N.R., Chen, S. (2000). *Power System Quality Assessment*. New York: John Wiley & Sons.
4. IEEE standard dictionary of electrical and electronic terms, IEEE Standard 100, 1984.
5. IEEE recommended practices and requirements for harmonic control in electrical power systems, IEEE Standard 519, 1992.
6. Subjak J. S., Mcquilin, J. S. (Nov./Dec. 1990). Harmonics—causes and effects, measurements and analysis: an update. *IEEE Trans. on Industry Applications* 26(6).
7. Akagi, H. (1992). Trends in active power line conditioners. Vol. 1. In: Proc. IEEE Industrial Electronics, Control, Instrumentation, and Automation, pp. 19–24.

8.   Frank T. M., Divan, D. M. (Sep/Oct. 1998). Active filter system implementation. *IEEE Industry Application Magazine* 4.

9.   Martzloff, F., Gruzs, T. (Nov./Dec. 1990). Power quality site surveys: facts, fiction and fallacies. *IEEE Trans. on Industry Applications* 26(6).

10. IEEE Working Group on Nonsinusoidal Situations (Jan. 1996). A survey of north American electric utility concerns regarding nonsinusoidal waveforms. *IEEE Trans. Power Delivery* 11(1).

11. Recommended practice for establishing transformer capability when supplying nonsinusoidal load currents, IEEE Std. C57.110-1998, March 1999.

12. Wei H., Batarseh, I. (1999). Comparison of basic converter technology for power factor correction. In: Proc. IEEE Southeast Conf., pp. 348–353.

13. Singh, B., Al-Haddad, K., Chandra, A. (Oct. 1999). A review of active filters for power quality improvement. *IEEE Trans. on Industrial Electronics* 46(5).

14. Shimamura, T., Kurosawa, R., Hirano, M., Uchino, H. (1989). Parallel operation of active and passive filters for variable speed cycloconverter drive systems. Vol. 1. In: Proc. 15th IEEE Industrial Electronics Society Conf., pp. 186–191.

15. Benchaits I., Saadate, S. (1996). Current harmonic filtering of non-conventional non-linear load by current source active filter. Vol. 2. In: Proc. IEEE International Symposium on Industrial Electronics, pp. 636–641.

16. Pottker de Souza, F., Barbi, I. (1999). Power factor correction of linear and nonlinear loads employing a single phase active power filter based on a full-bridge current source inverter controlled through the sensor of the AC mains current. Vol. 1. In: Proc. 30th IEEE Conf. on Power Electronics Specialists, pp. 387–392.

17. Buso, S., Malesani, L., Mattavelli, P. (Oct. 1998). Comparison of current control techniques for active filter applications. *IEEE Trans. on power electronics Industrial Electronics* 45(5):722–729.

18. Benchaita, L., Saadate, S., Salem nia, A. (May 1999). A comparison of voltage source and current source shunt active filter by simulation and experimentation. *IEEE Trans. on power electronics* 14(2):642–647.

19. Marks J. H., Green, T. C. (2001). Ratings analysis of active

power filters. Vol. 3. In: Proc. 32nd IEEE Power Electronics Specialists Conference, pp. 1420–1425.

20. Akagi, H. (2000). Active and hybrid filters for power conditioning. Vol. 1. In: Proc. IEEE Conf. on Industrial Electronics, pp. TU26–TU36.

21. Peng, F. Z. (July/Aug. 2001). Harmonic sources and filtering approaches. *IEEE Industry Applications Magazine* 7(4):18–25.

22. Fujita, H., Akagi, H. (March 1998). The unified power quality conditioner: the integration of series- and shunt-active filters. *IEEE Trans. on Power Electronics* 13(2):315–322.

23. Walker, J. (April 1984). Designing practical and effective active EMI filters. In: Proc. IEEE 11th Power Conf., Paper I3.

24. Farkas, T., Schlecht, M. F. (May 1994). Viability of active EMI filters for utility applications. *IEEE Trans. on Power Electronics* 9(3):328–337.

Index

*Note: Page references followed by f
denote figures and page references
followed by* t *denote tables.*